# PRACTICAL HYDROPONICS

# PRACTICAL HYDROPONICS

## *Handbook for Home Gardeners*

STEVEN CARRUTHERS

BLUEWATER
PRESS

Published by Bluewater Press
Warilla, New South Wales, Australia

A catalogue record for this book is available from the National Library of Australia.

ISBN 978-1-7647464-0-3 (paperback)

Cover design: Steven Carruthers
Typeset in Adobe Garamond Pro and Bodoni using QuarkXPress 2025

Printed on demand in Australia, the United States and the United Kingdom

First Edition

# Contents

# Introduction

Hydroponics is the practice of growing plants without soil, using water enriched with the nutrients essential for plant growth. Whether supported by an inert medium or suspended directly in solution, the principle remains the same: precise control over water, nutrients and growing conditions.

The word itself is derived from the Greek *hydro* (water) and *ponos* (labour), a fitting description of a method that replaces the uncertainties of soil with a more controlled and efficient system. Today, hydroponics is often referred to as "soilless culture," though the concept extends beyond that simple definition.

In practice, the terms hydroponics and soilless culture are used interchangeably and are treated as such in this book. The defining principle is not the absence of soil, but the delivery of a complete nutrient solution directly to the plant. Whether that solution is applied in water alone or through an inert growing medium is secondary to this central concept.

While often regarded as a modern innovation, hydroponics has deep historical roots. The underlying principle—separating plant growth from natural soil systems and controlling the delivery of water and nutrients—can be traced back thousands of years.

*An artist's impression of the legendary terraced gardens, celebrated in antiquity as one of the Seven Wonders of the Ancient World.*

One of the earliest and most enduring examples is the legendary Hanging Gardens, long associated with Babylon but now widely attributed to the Assyrian king Sennacherib at Nineveh. These elevated gardens, sustained by engineered water systems of canals and aqueducts, demonstrate an advanced understanding of controlled cultivation. Although soil was used, the deliberate management of water and growing conditions reflects principles closely aligned with modern hydroponic practice.

*An Aztec chinampa food-production system, where canals served both irrigation and transport routes, wide enough for two boats to pass side by side.*

Comparable approaches appeared elsewhere. In Kashmir, floating gardens on Dal Lake have supported cultivation for centuries, with plants grown on buoyant organic beds supplied continuously with water. In Central America, the Aztecs constructed chinampas—artificial islands in shallow lakes—that enabled highly productive agriculture through constant access to moisture and nutrients. While not hydroponic in the modern sense, these systems represent early efforts to control the growing environment in order to increase reliability and yield.

Modern hydroponics emerged in the early twentieth century through the work of researchers in the United States and Europe who demonstrated that plants could be grown to maturity using nutrient solutions in controlled environments. Among them, Dr William F. Gericke of the University of California played a prominent role in bringing hydroponics to public attention, particularly through his promotion of large-scale water culture systems.

While Gericke's work helped establish hydroponics as a practical method rather than a purely experimental concept, it formed part of a broader body of research undertaken by a number of scientists working with both solution and media-based systems. These parallel developments were critical in shaping the commercial and practical application of hydroponics as it evolved through the twentieth century.

During the Second World War, hydroponics proved its operational value. Allied forces established growing systems on remote islands and military outposts where fresh produce was otherwise unavailable, confirming that food production could be sustained in environments unsuitable for conventional agriculture.

*Workers tending hydroponic vegetable beds at Camp Chofu, Japan, developed to supply fresh produce to United States occupation forces and later expanded during the Korean War.*

Since then, hydroponics has developed into a widely adopted method of cultivation. Advances in materials, nutrient chemistry and environmental control have enabled efficient production in greenhouses, urban settings and regions with poor or degraded soils. In an era of increasing pressure on land and water resources, hydroponics represents a practical and scalable alternative.

For the home gardener, hydroponics provides a productive and space-efficient way to grow fresh vegetables, herbs and flowers. Because nutrients are delivered directly to the roots, plants grow more efficiently and can often be harvested sooner than those grown in soil. Water use is reduced, and many common soil-related problems—such as weeds, pests and diseases—are minimised.

*Hydroponic gravel beds at Camp Chofu used crushed volcanic pumice as the growing medium for large-scale vegetable production.*

There is no fundamental difference in how plants function, whether grown in soil or in a hydroponic system. The essential processes of nutrient uptake and growth remain the same. The difference lies in how those nutrients are delivered. In soil, nutrients must be broken down from organic matter before becoming available to plants. In hydroponics, they are supplied in a readily accessible form, allowing for more consistent and controlled growth.

Hydroponics is not a replacement for traditional gardening, but an extension of it. The same basic principles apply—light, water, nutrients and care—yet the method offers greater control and flexibility. Whether used on a commercial scale or in a small backyard system, hydroponics continues to expand the possibilities of how and where we grow our food.

This book is intended as a practical guide to those possibilities. It draws on both established principles and modern developments to provide a clear pathway for anyone wishing to design, build and maintain a hydroponic system with confidence.

## Why Grow Hydroponically?

Hydroponics offers practical advantages where space, water and soil quality are limiting factors. Its strength lies not in replacing traditional gardening, but in providing a controlled alternative where conventional methods are less effective.

Because nutrients are delivered directly to the root zone in a balanced solution, plants grow with a high degree of consistency. This allows for more predictable results and, in many cases, shorter growing cycles. The efficiency of nutrient delivery also enables closer planting, making better use of available space.

Water use is carefully managed. In most systems, water is recirculated and reused, reducing overall consumption compared to soil-based gardening. This makes hydroponics particularly suited to areas where water is limited or where efficient use of resources is a priority.

Hydroponic systems also reduce many of the constraints associated with soil. Weeds are eliminated, and soilborne pests and diseases are largely avoided. As a result, plant management becomes more controlled, and reliance on chemical treatments can often be reduced.

Another significant advantage is flexibility. Hydroponic systems can be established in locations where traditional gardening is impractical—on balconies, in courtyards, indoors, or in areas with poor or contaminated soil. This adaptability has contributed to its increasing use in urban environments and small-scale applications.

These advantages are best understood in practical terms: hydroponics provides a method of cultivation that is efficient, controlled and adaptable. For the home gardener, this translates into reliable production, better use of space and resources, and the ability to grow fresh produce in a wide range of environments.

## Understanding the Limitations

While hydroponics offers clear advantages, it also requires a more deliberate approach than traditional soil gardening. Success depends on consistency, attention to detail and an understanding of how the system functions as a whole.

Because plants rely entirely on the nutrient solution provided, any imbalance or interruption can affect growth more quickly than it would in soil. For this reason, hydroponic systems benefit from regular monitoring and timely adjustment to maintain stable growing conditions.

There is also an initial investment involved in setting up a system. Simple arrangements can be constructed from readily available materials, while more advanced systems may incorporate pumps, reservoirs and specialised growing containers. The level of complexity depends largely on the needs of the grower.

Plant health must be observed closely. In recirculating systems, problems can spread through the nutrient solution if not identified early. With sound management and good hygiene practices, however, these risks can be effectively controlled.

Hydroponics also relies on a consistent supply of water and nutrients. Unlike soil, which can buffer short-term fluctuations, hydroponic systems operate within narrower margins and require a more structured approach to maintenance.

These considerations do not represent barriers, but part of the discipline of hydroponic growing. With a basic understanding of the system and regular attention, they become routine aspects of successful cultivation.

## A Balanced Approach

For most gardeners, the advantages of hydroponics outweigh its limitations once the basic principles are understood. Like any method of cultivation, success depends on knowledge, observation and consistency in practice.

Hydroponics should be regarded not as a replacement for traditional gardening, but as an extension of it. The same fundamental requirements apply—light, water, nutrients and care—yet the method offers a higher degree of control over how those elements are delivered.

The practical value of hydroponics lies in that control. When properly managed, it enables efficient use of space and resources, reliable plant growth, and the ability to cultivate in environments where soil-based methods are less effective.

The following chapters build on these principles, providing a structured guide to the systems, materials and techniques required to establish and maintain a successful hydroponic garden.

## What You'll Learn in This Book

This book is designed as a practical guide to hydroponic gardening, suitable for both beginners and those seeking to refine their existing knowledge.

It explains how hydroponic systems function, including the roles of water, nutrients and growing media in supporting plant growth. The principles of plant nutrition are presented in clear terms, enabling an understanding not only of what to do, but why it works.

A range of hydroponic systems is examined, from simple, low-cost setups suitable for home use through to more advanced arrangements. Step-by-step guidance is provided to assist in constructing and managing a system using readily available materials.

The book also addresses plant selection, preparation of nutrient solutions, and the monitoring of plant health. Common problems are identified, along with practical methods for maintaining consistent and reliable results.

In addition, it considers how hydroponic techniques can be adapted to different environments, including small urban spaces, indoor growing areas and locations where soil conditions are poor or unsuitable.

By the end of the book, the reader will have the knowledge required to design, build and maintain a hydroponic system with confidence, and to produce fresh, high-quality plants efficiently and sustainably.

# Plant Nutrition

Modern hydroponics is founded on an understanding of plant nutrition. Early research into plant constituents led to the identification of essential elements, establishing the basis for controlled cultivation.

In hydroponic systems, nutrient management is central to success. The composition of the nutrient solution determines plant growth, and must be maintained within appropriate limits throughout the growing cycle.

Hydroponics allows precise control over nutrient availability. Unlike soil, where nutrient levels are difficult to adjust once applied, the composition of a solution can be modified to suit each stage of development. This enables more consistent growth and more efficient use of inputs.

Because nutrients are supplied in a readily available form, plants are not required to compete for or search out essential elements. As a result, growth is more uniform and, in many cases, more rapid than in soil-based systems.

## Basic elements

Of the ninety-two naturally occurring mineral elements, only around sixty have been identified in plants. Of these, seventeen are considered essential for plant growth. To be classified as essential, an element must meet four criteria:

- It must be required for the plant to complete its life cycle.
- Its function must be specific; not replaceable by another element.
- It must be directly involved in plant nutrition.
- It must not counteract the toxic effects of another element.

These seventeen essential elements are divided into two groups: macro-elements, which are required in relatively large quantities, and micro-elements (or trace elements), which are needed in much smaller amounts.

The macro-elements are carbon (C), hydrogen (H), oxygen (O), nitrogen (N), phosphorus (P), potassium (K), calcium (Ca), magnesium (Mg) and sulphur (S). The micro-elements are iron (Fe), chlorine (Cl), boron (B), manganese (Mn), copper (Cu), zinc (Zn), molybdenum (Mo) and nickel (Ni).

**Elements**

Elements are listed in several groups or classifications, depending upon the amount normally used by plants in their growth and development.

**Macro-Elements**

Nitrogen (N)
Phosphorus (P)
Potassium (K)

**Secondary macro-elements**

Calcium (Ca)
Magnesium (Mg)
Sulphur (S)

**Micro-elements**

Iron (Fe)
Boron (B)
Zinc (Zn)
Copper (Cu)
Manganese (Mn)
Molybdenum (Mo)
Chlorine (Cl)
Nickel (Ni)

# Essential elements and what they do

## Macro-elements

**Carbon** (C) is a constituent found in all organic compounds.
**Hydrogen** (H) is a constituent of all organic compounds of which carbon is a constituent. It is important for the cation exchange in plant-medium relations.
**Oxygen** (O) is a constituent of many organic compounds. It is essential in the anion exchange between roots and the external medium.
**Nitrogen** (N) is used in various forms to promote rapid vegetative growth, leaf, flower, fruit and seed development, and chlorophyll development; and to increase the protein content in all plants.
**Phosphorus** (P) promotes and stimulates early growth and blooming and root growth. It hastens maturity and seed growth, and it contributes to the general hardiness of plants.
**Potassium** (K) promotes disease resistance and good development of carbohydrates, starches and sugars, and it increases fruit production.
**Calcium** (Ca) is vital in all parts of plants to promote the translocation of carbohydrates, healthy cell wall structure, strong stems, membrane maintenance and root structure development. It sometimes interferes with the ability of magnesium to activate enzymes.
**Magnesium** (Mg) promotes the absorption and translocation of phosphorous. It activates many enzymes and it appears to aid in the formation of oils and fats. It also plays a role in carbon dioxide assimilation.
**Sulphur** (S) promotes the synthesis of oils, good cell wall structure, and the synthesis and function of protein.

**Note**

As a general rule plants have a higher nutrient requirement during cooler months and a lower requirement in the warmer months. A stronger nutrient solution should therefore be maintained during winter and a weaker solution in summer when plants take up and transpire more water than nutrient.

## Micro-elements

**Iron** (Fe) acts as a catalyst in the photosynthesis and respiration process, and it is essential for the formation of sugars and starches. Iron also activates certain other enzymes.
**Chlorine** (Cl) is essential for photosynthesis where it acts as an enzyme activator during the production of oxygen from water.
**Boron** (B) is essential for cell wall strength and development, cell division, fruit and seed development, sugar transport and hormone development. Some functions of boron inter-relate with those of nitrogen, potassium, phosphorus and calcium in plants.

**Nickel – A Trace but Essential Nutrient**

Although nickel is recognised as the seventeenth essential plant nutrient, plants require only minute amounts. In most hydroponic systems, sufficient nickel is supplied naturally from water, fertilisers and metal components, so it is rarely included as a separate ingredient in nutrient formulations.

**Manganese** (**Mn**) activates one or more enzymes in fatty acid synthesis and the enzymes responsible for DNA and RNA formation. It also participates directly in the photosynthetic production of oxygen from water and may be involved in chlorophyll formation. Manganese is closely associated with copper and zinc.

**Copper** (**Cu**) is an internal catalyst and acts as an electron carrier. It is also thought to be involved in nitrogen fixation.

**Zinc** (**Zn**) with copper and manganese is linked to chlorophyll synthesis. It is also essential for auxin metabolism.

**Molybdenum** (**Mo**) acts as an electron carrier in the conversion of nitrate to ammonium. It is essential for nitrogen fixation and nitrate reduction.

*Single-part liquid nutrient solutions*

*Two-part liquid Bloom nutrient solutions*

*Two-part liquid Grow nutrient solutions*

## Nutrient solutions

In hydroponic systems, all essential elements are supplied in the form of a nutrient solution—fertiliser salts dissolved in water. While commercial growers often formulate solutions for specific crops, the home gardener can rely on a range of high-quality, general-purpose products that are pre-formulated and readily available in liquid or powder form.

These formulations remove much of the guesswork from plant nutrition. However, care should be taken to avoid products that contain sludge or dissolve poorly, as these indicate low-quality fertiliser salts.

Solubility is critical in hydroponics. Nutrient salts must remain fully dissolved in solution to be available to plant roots. Poorly soluble salts, or the presence of sludge, suggest impurities such as clay, silt or other inert materials that do not supply nutrients. These can reduce nutrient availability and may also block feeder lines in recirculating systems.

Most commercially available nutrient solutions contain all essential elements and are suitable for general use. Some are formulated specifically for particular growing media, such as rockwool, which tends to be more alkaline and therefore requires a different nutrient balance.

Nutrients are available in both liquid and powder forms. Liquid formulations are often preferred by home gardeners for their consistency and ease of mixing, while powders offer longer storage life and lower cost.

Many products are also supplied as separate ‘grow’ and ‘bloom’ formulations to match the changing needs of plants. During the vegetative stage, plants require higher levels of nitrogen to support leaf and stem development. As they mature and enter the flowering and fruiting phase, demand shifts toward potassium and phosphorus. Adjusting the formulation accordingly helps optimise growth and yield.

High-quality one-part nutrients are also available, often using chelated elements to improve nutrient availability. These are convenient and effective across a wide range of systems, though typically more expensive.

## Replacing the solution

Individual plants absorb nutrients at different rates and in varying proportions, so some elements become depleted sooner than others. Determining the exact nutrient balance of a hydroponic solution requires laboratory analysis and is impractical for most growers. The only reliable safeguard against nutrient disorders is therefore to replace the solution regularly. Spent solution should be disposed of responsibly, for example by diluting it with water and applying it to soil gardens or lawns. In most hydroponic systems, where the solution is recycled, its effective life is typically two to three weeks, depending on the season, crop type, and stage of plant growth. During the hotter summer months, when plants transpire more and take up greater volumes of water—thereby concentrating the remaining nutrients—the solution may need to be replaced as often as once a week, particularly when plants are well advanced in growth.

Even after a system has been drained, it will not return to a completely fresh starting point. Residual solution remains within pipework, channels, pumps, and growing media, and this can amount to a significant volume. When a new nutrient solution is introduced, it mixes with this residue, creating a temporary imbalance and contributing to the gradual accumulation of non-essential ions such as sodium. Over time, this build-up can interfere with nutrient uptake and overall plant performance. In smaller systems, it is therefore good practice to periodically flush the entire system with fresh water before refilling, helping to remove accumulated salts and restore a more balanced nutrient environment.

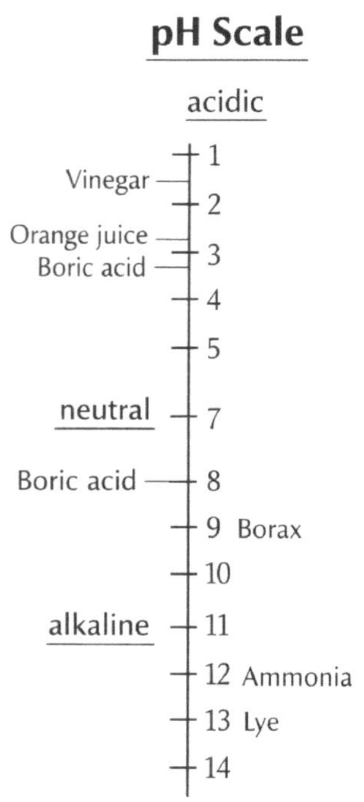


*Most crops grow best between pH 5.5 and 6.5*

# Nutrient management

While optimum nutrition is readily achieved in hydroponics, plants can be just as easily damaged through errors in preparing the nutrient solution or through failure to monitor and adjust it regularly. In systems that rely on automatic controls, poor maintenance or equipment failure can also lead to serious plant loss. The success or failure of a hydroponic garden therefore depends on a disciplined nutrient management programme, achieved through careful control of pH, temperature, and electrical conductivity (EC) of the solution. This control, combined with a sound working knowledge of these factors, is fundamental to successful hydroponic gardening

# pH level

In simple terms, pH is a measure of acidity or alkalinity on a scale of 1 to 14. The term is familiar from swimming pools and personal care products. In chemistry, pH is more precisely defined as the cologarithm of the activity of dissolved hydrogen ions (H+). Because hydrogen ion activity cannot be measured directly, pH values are based on theoretical calculations and standard reference solutions established by international agreement. The pH scale is therefore relative rather than absolute.

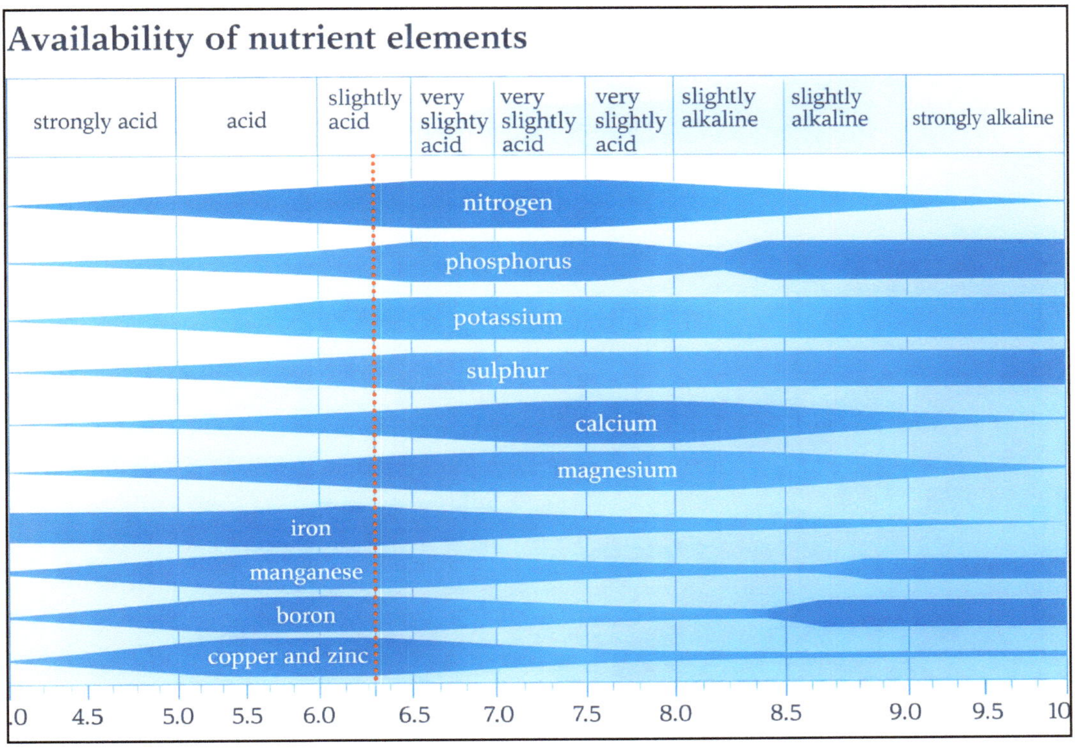


Nutrient deficiencies will become apparent if the pH rises above or falls below the recommended range for a given plant. For example, where the pH is consistently around 7.5, interveinal chlorosis is likely to occur, indicating reduced iron availability. The chart spans a pH range from 4.0 to 10.0, with the width of each shaded blue band representing the relative availability of that nutrient—widest at maximum availability and narrowing as availability declines. The dotted line at pH 6.25 marks the point at which the greatest number of nutrients are simultaneously available at or near their optimum levels.

## Nutrients & light

Low light conditions cause plants to take up more potassium and phosphorus, thereby making the nutrient solution more acidic. Conversely, on clear sunny days, plants take up more nitrogen, making the nutrient solution more alkaline.

**WARNING!**
**Take Care with Acids**

Concentrated acids are very dangerous, especially to the eyes, so always wear goggles, rubber gloves and protective clothing.

If acid gets onto someone, immediately flood the area with lots of water and continue for some time. Take particular care when diluting acid.

NEVER add water to concentrated acid, which will instantly boil and splatter. The safe technique is to have water in the container and stir while adding the acid slowly. Check the temperature and leave to cool if too hot.

In a nutrient solution, pH governs the availability of essential plant elements. A solution is neutral at pH 7.0, alkaline above that point, and acidic below it. Pure water at 25°C (77°F) is close to neutral, but even small changes in pH represent large changes in acidity or alkalinity. A solution at pH 6.0, for example, is ten times as acidic as one at pH 7.0, while a solution at pH 5.0 is 100 times as acidic.

In soil, most plants prefer conditions that are slightly acidic. Hydroponically grown plants generally favour a somewhat lower range, with an optimum pH usually between 5.8 and 6.5. Above pH 7.5, iron, manganese, copper, zinc and boron become less available. Below pH 6.0, the availability of phosphorus, calcium and manganese declines sharply. Although hydroponic crops can be grown outside the ideal range, the further the pH drifts from it, the greater the risk of nutritional disorders.

In soil, acidity or alkalinity is usually corrected by adding lime or dolomite. In hydroponics, however, nutrient solution pH is adjusted with chemical buffers. It can be lowered with dilute phosphoric, sulphuric or nitric acid, and raised with a dilute solution of potassium hydroxide. For convenience, ready-made pH adjustment solutions are widely available from hydroponic suppliers, retail garden centres and large hardware stores such as Bunnings.

In hydroponics, movement in pH can also provide a useful indication of plant activity. If the pH is rising and the solution is becoming more alkaline, the plants are generally taking up more acidic nutrient salts. If the pH is falling and the solution is becoming more acidic, they are taking up more alkaline components.

As a general rule, pH should be tested daily with either a simple water-testing kit or an electronic meter, and adjusted if necessary with acid or alkaline buffers. Ideally, this should be done at the same time each day and, if possible, at the same temperature, since pH can fluctuate significantly with changes in light and temperature. During daylight hours, intense photosynthetic activity tends to drive pH upward. At dusk, when photosynthesis ceases and respiration dominates, pH tends to fall. Trying to correct these short-term natural swings can send the solution onto a harmful roller coaster of repeated adjustment. Staying within the recommended range is important, but avoiding large fluctuations is even more so.

In practice, most plants will grow reasonably well within a pH range of about 5.0 to 7.0. This refers to the solution around the roots, not the feed. If you can afford one, buy a pH meter; if not, pH papers and strips are adequate. pH only becomes a problem if it stays outside this relatively wide range, especially well above pH 7.

*A nutrient solution should be tested daily for both pH and electrical conductivity (EC). Because plants differ in their nutrient requirements, crops should be grouped according to their average EC needs. Most plants will tolerate a relatively wide pH range and, in general, if the EC is correct, the pH will usually remain within acceptable limits.*

# Temperature

Temperature fluctuations in a hydroponic solution influence not only pH, but also the solubility and availability of nutrients. Studies indicate that the ideal solution temperature for maximum nutrient solubility lies between 20°C and 22°C. Outside this range, trace elements become less soluble, reducing uptake by plants and potentially affecting growth and yield.

Temperatures above or below the optimum can stress plants in much the same way as extremes of pH. However, most hydroponic crops will still perform reasonably well across a broader range, typically between 18°C and 28°C.

# Electrical conductivity

As plants grow, they selectively absorb the elements they require, continually altering the balance of the remaining nutrient solution. This process is monitored through electrical conductivity (EC), which provides a measure of the total concentration of dissolved salts in the solution.

To gauge nutrient uptake, the solution should be tested daily and adjusted where necessary. If readings indicate a high nutrient concentration, it suggests that plants are absorbing water more rapidly than nutrients. As water is removed, the solution volume decreases and nutrient strength rises, potentially to harmful levels. In this situation, fresh water should be added to dilute the solution back to the desired concentration.

Conversely, if nutrient concentrations fall below the recommended level, plants are taking up nutrients more rapidly than water, and the solution should be strengthened by adding additional nutrients.

A range of simple and inexpensive electronic meters allows the home gardener to measure electrical conductivity quickly and accurately. These instruments may display readings as EC (milliSiemens per centimetre, mS/cm), total dissolved solids (TDS, expressed in parts per million, ppm), or conductivity factor (cF). As a guide, 1 mS/cm is equivalent to a cF of 10, and approximately 650 ppm on a TDS scale. Conversely, ppm values can be converted to EC by dividing by 650.

For reliable results, a meter capable of measuring nutrient

## Conversion chart

| mS/cm | cF | ppm |
|---|---|---|
| 0.5 | 5 | 350 |
| 1.0 | 10 | 650 |
| 1.5 | 15 | 975 |
| 2.0 | 20 | 1300 |
| 2.5 | 25 | 1625 |
| 3.0 | 30 | 1950 |
| 3.5 | 35 | 2275 |
| 4.0 | 40 | 2600 |
| 4.5 | 45 | 2925 |
| 5.0 | 50 | 3250 |

strength is indispensable for both commercial growers and serious home gardeners. Where a meter is not available, the nutrient solution should be replaced regularly—typically on a weekly basis—to maintain a stable balance of essential elements.

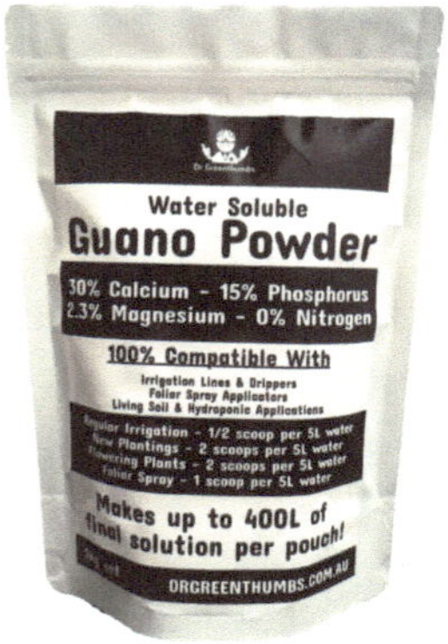

*A range of liquid and powdered organic fertilisers are available for hydroponic applications including bat and bird guano, and worm castings.*

## Plant requirements

Plants can generally be classified as low, medium or heavy feeders, and require correspondingly low, medium or high nutrient strengths. Only plants within the same category should be grown together using a common nutrient solution. Mixing different feeding types will favour one group while limiting the growth and development of others.

Plants grown outside their optimum EC or TDS range will produce inferior fruit and flowers. A high-strength solution, for example, can cause low feeders such as lettuce to develop a bitter taste. Conversely, a weak solution will result in medium feeders such as strawberries, and heavy feeders such as tomatoes, producing fruit that is soft, watery and lacking in flavour.

In the early development of hydroponics in Australia, when reliable data on plant nutrition was limited, hydroponically grown produce often gained a reputation for poor flavour. Today, with a better understanding of nutrient management, properly grown hydroponic fruit and vegetables are full of flavour. If produce lacks taste or vigour, it is almost always an indication that it has been grown outside the optimum nutrient range.

For recommended EC or TDS ranges for specific crops, refer to the Nutrient Guide on page 26-29.

## Organic and inorganic solutions

Hydroponics and organic growing have often been seen as unlikely companions. Traditionally, proponents of each have regarded the other with some scepticism, viewing the two approaches as fundamentally opposed. In practice, however, there is considerable common ground, and this overlap is attracting increasing interest among both new and experienced hydroponic growers.

Organic nutrient products are derived from natural sources or from combinations of natural materials. For those who prefer to avoid purely inorganic formulations, a range of organic nutrients has been developed specifically for hydroponic use. Common examples include products based on earthworm castings, seabird and bat guano, and seaweed extracts. These are available in powdered, granulated and water-soluble forms, and can be used either alone or as supplements to inorganic fertilisers.

Organic approaches aim to work with natural biological processes and are often credited with improving plant resilience, helping to reduce susceptibility to pests and diseases. However, it is important to recognise that hydroponically grown produce—when properly managed—is equal in nutritional value to that grown by organic or conventional methods, and is often superior in consistency and yield.

Inorganic fertilisers are derived from naturally occurring mineral sources that are refined into forms readily available for plant uptake. Because they are highly concentrated, relatively small quantities are required, making them efficient to transport, store and apply. They can also be precisely formulated to deliver the correct balance of nutrients for different stages of plant growth and fruiting.

By comparison, organic fertilisers—such as animal manures and plant residues—generally contain lower concentrations of nutrients and must be broken down before they become available to plants. This process can be slower and less predictable, and often requires larger volumes of material to achieve the same effect.

Both approaches have their place in modern hydroponics, and the choice ultimately depends on the grower's objectives, preferences, and level of management control.

# PLANT NUTRIENT GUIDE

## Herb Class

| HERB | pH | EC | ppm | Class |
|---|---|---|---|---|
| Basil | 5.5-6.5 | 1.8-2.2 | 1170-1430 | M |
| Chervil | 5.5-6.5 | 1.8-2.2 | 1170-1430 | M |
| Chives | 6.0-6.5 | 1.8-2.2 | 1170-1430 | M |
| Coriander | 6.0-6.5 | 1.8-2.2 | 1170-1430 | M |
| Dill | 6.0-6.5 | 1.8-2.2 | 1170-1430 | M |
| Fennel | 6.4-6.8 | 1.0-1.4 | 650-910 | L |
| Garlic | 6.0-6.5 | 1.4-1.8 | 910-1170 | M |
| Ginger | 5.5-6.5 | 1.0-1.2 | 650-780 | L |
| Lavender | 6.4-6.8 | 1.0-1.4 | 650-910 | L |
| Lemon balm | 5.5-6.5 | 1.6-2.4 | 1040-1560 | M |
| Marjoram | 5.5-6.5 | 1.6-2.0 | 1040-1300 | M |
| Mint | 6.5-7.0 | 1.8-2.2 | 1170-1430 | M |
| Mibuna | 6.0-6.5 | 0.8-1.2 | 520-780 | L |
| Mizuna | 6.0-6.5 | 0.8-1.2 | 520-780 | L |
| Mustard cress | 6.0-6.5 | 1.2-2.4 | 780-1560 | M |
| Oregano | 5.5-6.5 | 1.8-2.2 | 1170-1430 | M |
| Parsley | 5.5-6.0 | 0.8-1.8 | 520-1170 | M |
| Rosemary | 5.5-6.0 | 1.0-1.6 | 650-1040 | M |
| Sage | 5.5-6.5 | 1.0-1.8 | 650-1170 | M |
| Shallots | 6.0-6.5 | 1.8-2.2 | 1170-1430 | M |
| Thyme | 5.5-7.0 | 1.2-2.0 | 780-1300 | M |
| Watercress | 6.5-6.8 | 0.4-1.8 | 260-1170 | L |

### EC (electrical conductivity)

An indicator of the strength of a nutrient solution, as measured by an EC meter. This is the preferred term used worldwide. Its most common unit is milliSiemens per centimetre (mS/cm). Another unit is microSiemens per centimetre (uS/cm). A limitation of EC is that it indicates only the total concentration of a solution and not the individual nutrient components.

### pH and EC values

The pH and EC values shown in this table are provided as general ranges. Actual plant requirements will vary according to species and variety, stage of growth, and local conditions, including climate and season.

As a general guide, plants tend to require higher nutrient strengths in cooler conditions, and lower strengths in warmer conditions, when water uptake and transpiration exceed nutrient uptake.

A high pH can reduce the availability of iron, manganese, boron, copper, zinc and phosphorus to plants. A low pH can reduce the availability of potassium, sulphur, calcium, magnesium and phosphorus.

## Vegetable Class

| VEGETABLE | pH | EC | ppm | Class |
|---|---|---|---|---|
| Artichoke | 6.5-7.5 | 0.8-1.8 | 520-1170 | L |
| Asparagus | 6.0-6.8 | 1.4-1.8 | 910-1170 | L |
| Bean (common) | 6.0-6.5 | 2.0-4.0 | 1300-2600 | H |
| Bean (Broad) | 6.0-6.5 | 1.8-2.2 | 1170-1430 | M |
| Beetroot | 6.0-6.5 | 1.8-5.0 | 1260-3500 | H |
| Broccoli | 6.0-6.8 | 2.8-3.6 | 1820-2340 | H |
| Brussel sprout | 6.5-7.0 | 2.6-3.2 | 1690-2080 | H |
| Cabbage | 6.5-7.0 | 2.6-3.2 | 1690-2080 | H |
| Capsicum | 6.0-6.5 | 1.8-2.2 | 1170-1430 | M |
| Carrot | 6.0-6.3 | 1.6-2.0 | 1040-1300 | M |
| Cauliflower | 6.5-7.0 | 1.6-2.0 | 1040-1300 | M |
| Celery | 6.5-7.0 | 1.8-2.4 | 1170-1560 | M |
| Cucumber | 5.5-6.0 | 1.0-2.6 | 650-1690 | M |
| Eggplant | 5.8-6.2 | 2.4-3.6 | 1560-2340 | H |
| Endive | 5.5-6.0 | 0.8-1.6 | 520-1040 | L |
| Fodder | 6.0-6.5 | 1.0-1.8 | 650-1170 | M |
| Leek | 6.5-7.0 | 2.0-3.2 | 1300-2080 | H |
| Lettuce | 6.0-6.5 | 0.8-1.2 | 520-780 | L |
| Marrow | 5.6-6.0 | 2.0-2.6 | 1300-1690 | M |
| Okra | 6.0-6.5 | 2.0-2.8 | 1300-1820 | M |
| Onion | 6.0-7.0 | 1.4-1.8 | 910-1170 | L |
| Pak choi | 6.5-7.0 | 1.4-2.0 | 910-1300 | M |
| Parsnip | 6.0-6.5 | 1.8-2.0 | 1170-1300 | M |
| Pea | 6.0-7.0 | 0.8-1.8 | 520-1170 | L |
| Pepino | 6.0-6.5 | 2.0-5.0 | 1300-3250 | H |
| Potato | 5.0-6.0 | 2.0-2.6 | 1300-1690 | M |
| Pumpkin | 5.5-7.5 | 1.8-2.6 | 1170-1690 | M |
| Radish | 6.0-7.0 | 1.4-1.8 | 910-1170 | M |
| Spinach | 6.0-7.0 | 1.4-1.8 | 910-1170 | M |
| Silverbeet | 6.0-7.0 | 1.4-1.8 | 910-1170 | M |
| Squash | 6.0-6.5 | 1.2-1.6 | 780-1040 | M |
| Sweet Corn | 6.0-6.5 | 1.6-2.6 | 1040-1690 | M |
| Sweet Potato | 5.5-6.0 | 2.0-2.6 | 1300-1690 | M |
| Tat soi | 6.0-6.5 | 0.8-1.2 | 520-780 | L |
| Tomato (large) | 5.5-6.5 | 2.4-3.6 | 1560-2340 | H |
| Tomato (small) | 5.5-6.5 | 2.0-2.6 | 1300-1690 | M |
| Turnip | 6.0-6.5 | 1.8-2.4 | 1170-1560 | M |
| Ung tso | 6.0-6.5 | 1.6-1.8 | 1040-1170 | M |
| Witlof | 6.5-6.7 | 2.0-2.4 | 1300-1560 | M |
| Zucchini | 6.0-6.5 | 1.2-1.6 | 780-1040 | M |

| Low (L) | Medium (M) | High (H) |
|---|---|---|
| 0.6 mS/cm | 1.5-2.4 mS/cm | 2.4-5.0 mS/cm |

## Flower Class

| FLOWER | pH | EC | ppm | Class |
|---|---|---|---|---|
| African Violet | 6.0-7.0 | 1.2-1.6 | 780-1040 | L |
| Aloe | 5.5-6.5 | 2.0-2.4 | 1300-1560 | M |
| Anthurium | 5.0-6.0 | 2.0-2.6 | 1300-1690 | M |
| Aster | 6.0-6.5 | 1.8-2.6 | 1170-1690 | M |
| Begonia | 6.0-6.5 | 1.4-1.8 | 700-900 | L |
| Bromeliad | 5.0-7.5 | 0.8-1.0 | 520-650 | L |
| Cacti | 6.0-6.5 | 1.2-1.8 | 780-1170 | L |
| Caladium | 6.0-7.5 | 1.6-2.0 | 1040-1300 | L |
| Carnation | 6.0-6.5 | 1.0-1.2 | 650-780 | L |
| Chrysanthemum | 6.0-6.5 | 1.8-2.4 | 1170-1560 | M |
| Citronella | 5.5-6.5 | 2.0-2.4 | 1300-1560 | M |
| Cyclamen | 5.5-6.5 | 2.0-2.4 | 1300-1560 | M |
| Cymbidium | 5.5-6.0 | 0.6-0.8 | 390-520 | L |
| Dahlia | 6.0-7.0 | 1.4-2.0 | 700-1300 | L |
| Dieffenbachia | 5.0-6.0 | 1.8-2.4 | 1170-1560 | M |
| Dracaena | 5.0-6.0 | 1.8-2.4 | 1170-1560 | M |
| Ferns | 6.0-6.5 | 1.6-2.0 | 1040-1300 | L |
| Ficus | 5.5-6.0 | 2.0-2.6 | 1300-1690 | M |
| Freesia | 6.5-7.2 | 1.0-2.0 | 650-1300 | L |
| Gypsophila | 5.0-6.5 | 1.6-2.0 | 1040-1300 | M |
| Geranium | 6.0-6.5 | 1.2-1.8 | 780-1170 | L |
| Gerbera | 5.5-6.5 | 1.6-2.0 | 1040-1300 | M |
| Gladiolus | 5.5-6.5 | 1.6-2.4 | 1040-1560 | M |
| Impatien | 5.5-6.5 | 1.8-2.0 | 1170-1300 | M |
| Kangaroo paw | 5.5-6.5 | 2.0-2.4 | 1300-1560 | M |
| Limonium | 5.0-6.5 | 1.6-2.0 | 1040-1300 | M |
| Marigold | 6.0-6.5 | 1.8-2.4 | 1170-1560 | M |
| Monstera | 6.0-6.5 | 1.2-1.8 | 780-1170 | M |
| Palms | 6.0-7.5 | 1.6-2.0 | 1040-1300 | M |
| Pansies | 6.0-6.5 | 1.8-2.4 | 1170-1560 | M |
| Poinsettia | 6.0-6.5 | 1.2-1.8 | 780-1170 | M |
| Rose | 5.5-6.0 | 1.8-2.2 | 1170-1430 | M |
| Snapdragons | 6.0-6.5 | 1.0-1.2 | 650-780 | L |
| Spathiphyllum | 6.0-6.5 | 1.2-1.8 | 780-1170 | M |
| Stock | 6.0-7.0 | 1.6-2.0 | 1040-1300 | M |
| Sweet pea | 6.0-6.5 | 1.8-2.4 | 1170-1560 | M |

## Fruit Class

| FRUIT | pH | EC | ppm | Class |
|---|---|---|---|---|
| Banana | 5.5-6.5 | 1.8-2.2 | 1170-1430 | M |
| Blueberry | 4.0-5.0 | 2.0-3.6 | 1300-2340 | H |
| Cherries | 6.0-7.2 | 2.2-3.2 | 1430-2080 | H |
| Currant | 6.0-6.5 | 1.4-1.8 | 910-1170 | M |
| Mandarin | 5.5-6.5 | 2.0-2.4 | 1300-1560 | M |
| Mango | 6.0-6.5 | 1.2-1.8 | 780-1170 | M |
| Melon | 5.5-6.0 | 2.0-2.6 | 1300-1690 | M |
| Passionfruit | 6.0-6.5 | 1.6-2.4 | 1040-1560 | M |
| Paw paw | 6.0-6.5 | 1.6-2.4 | 1040-1560 | M |
| Peach | 6.8-7.2 | 2.2-3.2 | 1430-2080 | H |
| Pineapple | 5.5-6.0 | 2.0-2.4 | 1300-1560 | M |
| Red apples | 6.8-7.2 | 2.2-3.2 | 1430-2080 | H |
| Rhubarb | 5.5-6.0 | 1.6-2.0 | 1040-1300 | M |
| Strawberry | 6.0-6.5 | 1.4-2.0 | 910-1300 | M |
| Watermelon | 5.8-6.2 | 1.6-2.4 | 1040-1560 | M |

# Note

## Total Dissolved Solids = TDS

TDS is the total mass of inorganic dissolved solids in a solution, expressed as ppm or mg/L. There are also meters that imply that they measure TDS in ppm. Unfortunately, this is not strictly correct, because they are actually giving converted EC readings and the conversion factor is often in error.

## Conductivity Factor (cF)

Conductivity Factor (cF) as measured by a cF meter is an old unit of measurement that uses a multification factor of 10 (i.e 1.0 mS/cm = cF 10). This term is used in the UK, NZ, and by some growers in Australia.

## Nutrient strength

For most plants the total concentration of nutrient elements in a solution should be between 500 and 1500 ppm so that osmotic pressure will facilitate absorption. However, some plant varieties, such as large tomatoes, may require a nutrient concentration well over 2000 ppm during fruiting. Lower values are preferred by low-feeding plants such as watercress.

# Growing media and hydroponic techniques

One of the first steps in establishing a hydroponic system is selecting a suitable growing medium. There are several options, each with advantages and limitations, so it is worth understanding their characteristics before deciding which best suits your system.

Many growers find that combinations of different media work well, although achieving the right balance often involves some trial and error. Whatever medium is chosen, it must suit both the crop and the type of hydroponic system being used.

In soilless culture, the growing medium—or substrate—does more than simply support the plant. For best results, it must also hold moisture effectively, provide good drainage, and maintain adequate oxygen around the roots.

The type of medium used also influences how nutrient solution is applied. Media with good capillary action can draw solution upward from below, allowing subirrigation. Media with little capillary capacity usually require surface watering through trickle, drip or spray systems.

*A cross section of plants growing in a combined gravel and sand mixture.*

Ideally, hydroponic media should be inert substrates that neither contribute nutrients nor alter the nutrient balance, giving the grower full control over plant nutrition. Hydroponic systems are broadly classified according to the type of medium used, including nutrient film technique (NFT), aggregate culture, rockwool, coco peat, sawdust and aeroponics.

Aggregate media include a wide range of materials such as crushed granite, river pebbles, brick shards, volcanic cinders, blue metal gravel and polystyrene mulch, as well as more commonly used products such as scoria, vermiculite, perlite and expanded clay. The choice depends on availability, cost and system design.

Moisture retention is one of the most important characteristics of any medium. It is influenced by the size, shape and porosity of the particles. Water is held both on the particle surfaces and within pore spaces. The more porous the material, the more water it can store. Smaller particles generally retain more water than larger ones because they have greater surface area and pack more closely.

However, moisture retention is only part of the equation. Good drainage is equally important. A highly retentive medium is useful only if excess water can drain away freely, allowing adequate aeration and oxygen movement through the root zone. Because roots need both water and air, many growers achieve the best results by blending media of different particle sizes. After watering, a good medium should contain about 10 to 20 per cent air and 35 to 50 per cent water by volume.

Another important characteristic is cation exchange capacity (CEC). Many essential nutrients—including potassium, calcium, magnesium and iron—occur in solution as positively charged ions, or cations. Media with negatively charged surfaces can attract and hold these nutrients, reducing leaching and keeping them available to plant roots for longer.

Media with a high CEC generally require less frequent nutrient application than those with a low CEC. Peat and vermiculite, for example, have relatively high CEC, while sand, perlite and polystyrene mulch have low CEC ratings.

Finally, hydroponic media must be free of toxic substances and salts, and should not significantly alter the pH of the nutrient solution. They must also be durable enough to retain their structure over time. Media that break down or compact will restrict drainage and aeration, ultimately reducing root health and plant performance.

**Cation exchange**

The ability of a growing medium to hold various nutrients, including ammonium-nitrogen, potassium, calcium, magnesium, iron, manganese, zinc and copper. A root medium with low cation exchange capacity does not retain nutrients well and so must be fertilised often.

**Recirculating systems**

Systems in which the nutrient solution is recycled and re-used. These are also known as 'closed' systems. Different systems can be either run continuously, as with NFT, or intermittently, as with the flood-and-drain technique..

**Non-recirculating systems**

Systems in which the growing medium is fed with fresh nutrient solution to replace what the plant has used, with a slight excess displaced that is not recycled through the system. Also known as 'open' or 'free-drainage' systems.

# Water culture

In a typical nutrient film technique (NFT) system, plants are grown in channels or gullies so that their roots are exposed to a shallow, continuously flowing film of nutrient solution, usually about 2 mm deep. Most NFT systems are recirculating: the solution flows past the roots, drains back to a reservoir, and is pumped through the system again.

NFT was originally developed by Dr Allen Cooper of Sussex, England, and remains one of the most efficient forms of water-based hydroponics. It is officially defined as a system in which 'a very shallow stream of water containing all the dissolved nutrients required for growth is recirculated past the bare roots of crop plants in a watertight gully'. The key principle is that the nutrient stream should remain shallow enough to keep the upper part of the root mass moist but exposed to air. This provides roots with a constant supply of both nutrients and oxygen.

Although the basic NFT concept has changed little since its development, successful operation depends on matching the system to the crop. Lettuce, for example, requires only a relatively low nutrient concentration and, because of its compact root system, is ideally suited to NFT. Tomatoes require a stronger nutrient solution and larger channels to accommodate their more vigorous root growth. Crops with extensive root systems can eventually restrict nutrient flow if the channels are undersized.

NFT systems are widely used for lettuce, Asian greens, Mediterranean herbs, endive, cucumber, zucchini, capsicum, strawberries, silverbeet and many other leafy vegetables and herbs. They are not suitable for root or tuber crops.

The principal advantage of NFT is that it provides roots with a near-ideal balance of water, oxygen and nutrients. In many other hydroponic systems, excess water can reduce oxygen availability, or poor aeration can limit nutrient uptake. NFT overcomes this by delivering all three simultaneously, making it capable of producing high yields and high-quality crops from relatively dense plantings.

Its main weakness is its dependence on sound design and uninterrupted operation. Poorly designed channels, incorrect fall, inadequate flow rates or poor maintenance can result in flooding, waterlogging and crop loss. NFT also relies heavily on a continuous supply of water and power. If a pump fails or power is lost and no back-up is available, plants can deteriorate quickly. Unlike media-based systems, there is little buffering capacity to protect the crop during a breakdown.

**NFT (nutrient film technique)**
A water-based system where nutrient solution flows down channels or gullies and is recirculated. A basic principle of the technique is that the nutrient solution should be maintained as a thin film to enable adequate oxygenation of the solution.

*Leafy vegetables growing in a nutrient film technique (NFT) system. A shallow, continuously flowing nutrient stream supplies water, nutrients and oxygen directly to the plant roots.*

# Aggregate culture

## Sand

Sand is one of the oldest hydroponic growing media, and it remains widely used because of its low cost, availability and reliability. However, not all sand is suitable. Granitic or silica-type sands are preferred, while calcareous sands should be avoided because they are too alkaline. Beach sand is generally unsuitable because of its high salt content, although some coastal growers have used it successfully after thorough washing and leaching. In a home garden, any washed beach sand should be chemically tested before use.

Sand used in hydroponics should not be excessively fine, as this can reduce aeration and cause puddling, often seen when water rises to the surface after vibration. The ideal aggregate is clean river sand, washed free of silt and clay. A particle size between 0.6 mm and 2 mm is generally suitable, allowing good drainage while retaining adequate moisture. Some growers prefer blended grades to improve structure. Coarser sands may drain more freely in cool, damp climates, but they can dry out too quickly in hotter regions.

Sand culture has a long history. Trials were conducted as early as the mid-nineteenth century, and one of the first published attempts to develop hydroponics commercially in the United States in 1929 used sand as the growing medium. Today, sand culture remains well suited to arid and semi-arid regions such as the Middle East, North Africa and many parts of Australia.

*A hydroponic sand bed growing mustard lettuce.*

The main advantages of sand culture are its low cost, simplicity and versatility. It can be used in either open (free-drainage) or closed systems. Sand has excellent capillary action, which promotes lateral movement of nutrient solution and helps maintain even moisture and nutrient distribution throughout the root zone. Its fine particles also provide good water retention, reducing the frequency of irrigation compared with some coarser media.

Unlike NFT and other water-based systems, sand culture offers some buffering in the event of power failure, pump breakdown or pipe damage. Plants can continue drawing on retained moisture for a period before stress occurs, giving the grower time to make repairs. Practical advantages include lower construction costs, straightforward operation, and relatively simple maintenance.

The main disadvantages are the need to sterilise the medium

between crops to control medium-borne diseases, and the risk of salt build-up over time. Sterilisation by steam or approved chemical treatment is effective, although it can be time-consuming. Salt accumulation can be controlled by periodic flushing and by monitoring drainage water. In free-drainage systems, nutrient loss is higher than in recirculating systems, but with good management, wastage should be limited.

Sand culture is normally irrigated using a drip or trickle system. Beds should be laid to a slight fall, about 1:400, to assist drainage. A 13 mm black polyethylene supply line is usually run along the bed, with micro 'spaghetti' lines or emitters positioned at regular intervals. In wider beds, separate supply lines may be needed along each row.

Irrigation is typically carried out two to five times daily, depending on plant maturity, weather and seasonal conditions. In free-drainage systems, excess drainage should be tested regularly for dissolved salts. If salt levels become excessive, the bed should be flushed thoroughly with fresh water.

Sand beds can be built on waist-high benches or directly on the ground, making them suitable for sites with poor natural soil, bedrock or stony surfaces. Ground beds should be lined with plastic, with an agricultural drainage pipe running along the centre to remove excess nutrient solution. This pipe should be surrounded by coarse gravel to prevent clogging. Beds are commonly about 400 mm deep.

Sand culture is particularly well suited to Australian conditions, especially in drier inland and rain-shadow regions. In areas subject to prolonged heavy rain, however, sand beds can become waterlogged unless protected. In such cases, simple clear plastic covers or tunnel shelters can provide effective protection.

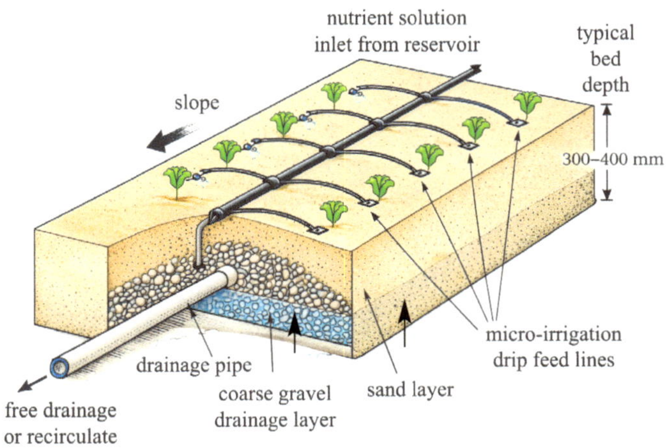


## Puddling

Puddling or water-logging is caused by a high percentage of silt and fine sand. Examples of puddling can often be seen in footsteps while walking along wet beach sand. When puddling occurs, plant roots are constantly immersed in nutrient solution, which can kill the plant.

## Gravel

Gravel shares many of the characteristics of sand, but its larger particle size gives it freer drainage and greater aeration, while reducing its water-holding capacity. A blend of sand and gravel can provide an excellent growing medium — ideally, about 40 per cent sand to 60 per cent medium-sized blue metal gravel, coarse river gravel or washed scoria.

Gravel culture is among the oldest and most widely used hydroponic techniques. It was one of the first methods employed by William F. Gericke, who helped establish modern soilless culture through sub-irrigation methods. It was also used by American servicemen on coral atolls in the South Pacific during the Second World War.

Many growers prefer gravel culture to water culture because the aggregate provides excellent root support. The gravel is placed in channel trays or containers similar to those used in water culture systems. Nutrient solution is stored in a separate reservoir and pumped through the aggregate as required to moisten the root zone. After irrigation, sufficient moisture and nutrients remain clinging to the aggregate particles to sustain plant roots until the next cycle.

As with sand, gravels of calcareous origin — such as limestone and coral — should be avoided, as they raise the alkalinity of the nutrient solution. This can make iron unavailable to plants and cause soluble phosphates to become insoluble.

*Snapdragons grown in a half-round open channel filled with medium-grade gravel.*

The best gravel for hydroponics is crushed granite or blue metal of irregular shape, free of fine particles smaller than 2 mm and coarse particles larger than 15 mm. Ideally, at least half the total volume should consist of particles around 10 mm in diameter. The gravel should be hard enough to resist breaking down over time, capable of holding moisture within its void spaces, and free-draining enough to ensure good aeration.

Other suitable aggregate media include vermiculite, perlite, expanded clay, scoria, versarock, zeolite, pumice, smooth river pebbles and crushed marble.

Gravel culture systems can be adapted to a range of hydroponic methods, including NFT channels filled with gravel and drip-irrigation systems.

## Vermiculite

Vermiculite is a flaky mineral with a structure similar to mica. It is mined in countries such as South Africa, China, Brazil, the United States and Australia. In its raw form, vermiculite consists of layered silicates of aluminium, iron and magnesium.

When heated to very high temperatures in a process known as exfoliation, the moisture trapped within the layers turns to steam, forcing the flakes apart. This causes the particles to expand dramatically — often by up to twenty times their original size — forming soft, lightweight, glittering granules ideal for hydroponic use.

*Vermiculite is widely used for seed raising, potting mixes and hydroponics, and is available in grades ranging from extra fine to coarse.*

One of vermiculite's main strengths is its ability to retain water, holding up to 50 per cent of its volume in moisture. This makes it especially useful in hydroponic systems where drying out may be a concern. In passive systems, it also has an excellent natural wicking ability, drawing water and nutrients through the medium to plant roots.Vermiculite has a near-neutral pH of about 7.0, although it can sometimes become slightly alkaline. Like perlite, it is sterile and free from harmful organisms. It also contains small amounts of magnesium and potassium, which can be gradually released to plants.

A further advantage is its relatively high cation exchange capacity and good buffering ability, meaning it can help stabilise nutrient availability and resist sudden pH changes.

Because vermiculite can retain too much moisture if used alone, it is often mixed with perlite to improve drainage and aeration. A blend of one part vermiculite to three parts perlite generally gives the best balance.

The main disadvantage of vermiculite is that it tends to break down over time. In recirculating systems this can lead to clogged tubing and filters, while in simpler systems it may reduce drainage and encourage stagnation.

## Perlite

Perlite is a volcanic mineral that, when mined, resembles small black crystals of sand. When heated to about 1000°C, the particles expand explosively — much like popcorn — to form soft white or grey granules with a lightweight, foam-like texture.

*Horticultural-grade perlite has many uses in the home and garden and is readily available from Amazon and specialist garden centres.*

Each expanded particle is honeycombed with tiny air pockets that can absorb water and nutrient solution, then release them gradually to plant roots. Although less spongy than vermiculite, perlite offers superior drainage, making the two media highly complementary when blended.

*Roses grown in Perlite.*

One of perlite's main advantages is its sterility. During manufacture, the material is heated to a point that eliminates disease organisms, making it a clean and reliable growing medium. Its light weight is also a major benefit, particularly for home gardeners, as it is easy to handle and work with.

Perlite has a relatively neutral pH, generally ranging from 6.5 to 7.7. It is particularly valued for improving drainage and maintaining aeration — both essential in hydroponic systems. It can be used on its own as a growing medium or mixed with vermiculite where greater water retention is needed, typically in a ratio of three parts perlite to one part vermiculite.

Unlike vermiculite, perlite has no significant cation exchange capacity or pH buffering ability, and it contributes no mineral nutrients of its own. However, its strong capillary attraction for water makes it highly effective in sub-surface irrigation systems.

Because of its rigid structure, perlite maintains excellent aeration over time and, if handled carefully, can often be reused for several successive crops.

## Expanded clay

Expanded clay, commonly known as LECA (light expanded clay aggregate), is produced by heating clay in rotary kilns, where it expands into small, lightweight, irregular pellets.

Also known as clay balls or hydro-pebbles, they are 100% inert, rot-resistant, and reusable.

*Expanded clay is popular with home gardeners for its excellent capillary properties and natural, earthy appearance.*

This medium is porous and offers excellent capillary action. Each pellet absorbs nutrient solution into its internal structure and delivers it efficiently to plant roots, while its light weight ensures good aeration throughout the root zone. The stable structure of expanded clay resists breakdown over time and helps prevent the build-up of acids, maintaining a consistently stable pH.

Because of its natural, earthy appearance, expanded clay is widely used for indoor potted plants as well as a range of simple hydroponic systems. However, in conditions where algae growth is a concern, its small pore spaces can become clogged.

Although expanded clay is imported into Australia, it is now more affordable than in the past and is increasingly popular in more advanced hydroponic systems — particularly flood-and-drain setups, where it has proven to be a highly effective growing medium.

## Scoria

Scoria is a hard, porous volcanic rock formed when gases become trapped in molten lava, leaving behind a structure full of air pockets. It is commonly basaltic or andesitic in composition. The name derives from the Greek *skoria*, meaning 'rust', reflecting its typical red colour, although it may also appear dark brown or black. An older term for scoria is cinder. It is found worldwide in regions of volcanic activity, with a notable Australian source at Mount Quincan in Far North Queensland.

Scoria is widely used in landscape design, where its colour and texture provide a natural, earthy appearance. As with other crushed rock media, it may be slightly alkaline (typically pH 7 to 10) and can contain soluble salts. For this reason, it should be thoroughly washed before use in hydroponics, and the nutrient solution may require closer monitoring and adjustment during the initial period.

Scoria can be crushed and graded to various particle sizes. Finer grades, in the range of 0.5 to 2 mm, behave in a similar manner to perlite, although scoria has a higher bulk density.

*Scoria is an economical hydroponic growing medium and is particularly well suited to orchids.*

## Zeolite

Zeolite is a relatively uncommon hydroponic medium in Australia. It is a naturally occurring mineral widely used as a soil additive due to its high cation exchange capacity — the ability to hold and exchange plant nutrients efficiently.

Structurally, zeolite has a distinctive cage-like molecular framework that enables it to absorb, store and release nutrients as required by plant roots. It is composed primarily of naturally occurring silicates, with deposits found worldwide. Different types vary in hardness (and therefore durability) as well as in their mineral composition.

Australian zeolites are typically rich in calcium, with moderate levels of available potassium. The material has a gravel-like appearance and is relatively heavy compared with other hydroponic media.

Some research in the United States has explored the use of zeolite in combination with other minerals, such as apatite, as a hydroponic growing medium, taking advantage of its nutrient-holding properties.

Unlike organic amendments, zeolite does not break down or decompose over time, providing long-term structural benefits for hydroponic gardens.

*Horticultural zeolite acts as a "nutrient reservoir," retaining fertilisers for plant use and reducing leaching, particularly in sandy soils.*

*Horticultural zeolite acts as a "nutrient reservoir," retaining fertilisers for plant use and reducing leaching, particularly in sandy soils.*

*Pumice is a light-coloured volcanic rock formed when superheated, pressurised magma is violently ejected and rapidly cooled.*

## Pumice

Pumice, like perlite, is a siliceous volcanic rock formed when gases become trapped in cooling magma, creating a lightweight, porous structure. Unlike perlite, it does not require heat expansion during processing; it is simply crushed and screened to the desired particle size.

Pumice is heavier than perlite and does not absorb water as readily, but it still provides good moisture retention — more so than gravel of a similar size. It is pH neutral and has a low cation exchange capacity, making it chemically stable and predictable in use.

Although often blended with materials such as peat and sand, pumice also performs well as a stand-alone hydroponic medium, offering a balance of drainage and aeration.

Widely available in New Zealand, pumice is less common in Australia due to the absence of significant local deposits and the high cost of transport.

## Plastic foam

Plastic foam media are available in pellet form and are both chemically inert and extremely lightweight. Although several types of plastic are used, these materials generally have poor water and nutrient retention and, when used alone, may not provide adequate support for plant roots.

They can be blended with other growing media to improve performance, but some types tend to float and separate over time. Plastic foams contain no nutrients, have negligible cation exchange capacity and offer little buffering ability.

Often referred to as plastic mulch, these pellets are typically made from materials such as polyurethane, polystyrene or urea-formaldehyde.

*Sphagnum moss is often used in passive hydroponics, hanging baskets, and as a growing medium for plants such as aroids.*

## Sphagnum moss

Sphagnum moss is a highly effective organic growing medium, valued for its exceptional moisture retention and suitability for propagation, carnivorous plants and orchids. It combines a high water-holding capacity with good aeration and a naturally acidic pH.

Capable of holding up to ten times its dry weight in water, sphagnum moss creates a sponge-like environment that helps prevent plants from drying out while maintaining adequate oxygen around the roots. It is commonly used on its own for propagation or blended with perlite to improve aeration, offering an alternative to more traditional inert media.

However, sphagnum moss is not a sustainable resource and decomposes over time. For this reason, it is not well suited to high-flow hydroponic systems, such as NFT, where the material can break down and impede water movement.

**Rockwool culture**
In wrapped rockwool installations the drainage holes must be cut between the drippers and at both side of the rockwool slab. Ensure that the outlets under the slabs are remote from drippers so that the whole slab can be used and excess salts leached away.

## Rockwool

Rockwool, developed in Denmark in 1969, is one of the most widely used hydroponic growing media in commercial horticulture worldwide. It was introduced to Australia in 1982 with the development of Growool, a horticultural-grade form of the material.

Rockwool is an inert, fibrous substrate manufactured from a blend of volcanic rock, limestone and coke, melted at temperatures of around 1500°C to 2000°C. The molten material is spun into fine fibres, then compressed into sheets, cubes or slabs. A wetting agent is added during manufacture to reduce surface tension and improve water absorption. It is important to note that insulation-grade rockwool used in the building industry is not suitable for horticultural use.

Horticultural rockwool is valued for its excellent balance of air and water retention, typically comprising about 3 per cent solid material and 97 per cent pore space. Root growth tends to follow the orientation of the fibres, which can be engineered to suit different stages of plant development. Propagation cubes are usually structured to encourage downward root growth, while slabs promote horizontal root expansion once plants are established.

As an inert medium, rockwool contains no nutrients and has negligible cation exchange capacity, meaning it does not retain or alter the nutrient solution. Plant nutrition is therefore entirely controlled by the composition of the solution supplied. Rockwool also has no buffering capacity and does not significantly influence pH after initial wetting. When first soaked, it may raise the pH of the solution slightly — typically by about one unit — but this can be corrected during conditioning. Thereafter, a feed pH in the range of 5.2 to 6.2 is generally suitable.

Before use, many growers condition rockwool by soaking it in nutrient solution for 24 to 48 hours to ensure even wetting and stabilise pH. Slabs should be placed in their final position during this process, with drip lines arranged along the top. Plastic wrapping should not be cut until the material is fully saturated.

*Rockwool is popular with home gardeners: seeds are started in small cubes, transferred to larger blocks, and then placed on slabs for final growth.*

Rockwool is available in a wide range of forms, from small propagation cubes to large slabs (commonly up to 1500 × 300 × 100 mm). Larger units are often plastic-wrapped to reduce evaporation and control root spread. Some cubes are designed with depressions to accept smaller propagation blocks, facilitating transplanting. Granulated rockwool, or "floc," is also available for pot culture or for improving aeration in mixes.

This medium performs well in both free-drainage and recirculating systems. In free-drainage systems, disease risk is reduced because nutrient solution is not reused. Its lightweight, self-contained nature allows for flexible plant spacing and efficient use of growing space — young plants can be raised at high density and later spaced out as they mature.

Rockwool slabs do not require perfectly level surfaces, as their plastic wrapping helps regulate internal water distribution. In tropical climates, however, the plastic may be removed to promote evaporative cooling of the root zone. Even then, separation between slabs is maintained to prevent uneven moisture distribution and disease transfer.

Irrigation is typically delivered via drip systems or, less commonly, NFT, with frequency determined by plant size and environmental conditions. In slab culture, drainage slots are cut low along the sides and ends to encourage even horizontal movement of the nutrient solution.

Water distribution within rockwool is not uniform: thicker slabs tend to hold more moisture at the base, while thinner slabs provide a better balance between water retention and aeration. For example, a 75 mm slab may retain around 77 per cent of its pore space as water, leaving adequate air-filled porosity for root health.

*Horticultural rockwool is a sterile, inert, and sustainable growing medium, ideal for hydroponic propagation and intensive plant cultivation.*

The advantages of rockwool include consistency, cleanliness, ease of handling and suitability for intensive production systems. It reduces labour requirements and eliminates much of the heavy preparation associated with traditional media.

Its disadvantages are relatively minor. Transport costs can be high due to its bulk, particularly in remote areas, although reuse can offset this. Handling dry rockwool may cause minor skin irritation, so gloves and long sleeves are recommended, and the material should be dampened before use.

Overall, rockwool remains one of the most versatile and reliable substrates available, particularly suited to modern hydroponic systems where precision control of water and nutrient delivery is essential.

## Coco peat

*Homrgrown tomotoes grown in coco peat slab.*

Coco peat — also known as coir pith, coir fibre or coir dust — is derived from coconut husks, a by-product of the coconut industry. The husks are washed, heat-treated and graded before being processed into a range of horticultural products of varying particle size and density.

In hydroponics, coco peat is used in systems similar to rockwool culture. It may be contained in plastic-wrapped slabs or used loose in pots and containers. It is commonly supplied as compressed bales, briquettes, slabs or discs, which expand significantly when water is added — one kilogram of coco peat can yield up to 15 litres of moist medium.

Coco peat has a cation exchange capacity comparable to sphagnum peat and offers excellent water retention while maintaining good aeration. It rewets readily after drying and provides a well-balanced root environment. The material also contains small amounts of naturally occurring nutrients, including calcium, magnesium and potassium.

As an organic medium, however, coco peat is not fully decomposed and may compete with plants for nitrogen — a process known as nitrogen drawdown — if nutrient levels are not adequately maintained. Lower-quality products may also contain excess sodium, so it is important to check the electrical conductivity (EC) of runoff and flush the medium if necessary. Good-quality coco peat is generally close to neutral pH.

Coco peat is suitable for use in recirculating, drip and flood-and-drain systems. For improved drainage, it can be blended with materials such as perlite, pumice or expanded clay.

Another advantage is its environmental profile: being organic, coco peat is easy to dispose of and is considered a more sustainable alternative to many inert media. With careful management, it can often be reused for up to three growing cycles with minimal loss of performance.

## Sawdust

Sawdust culture is used primarily in Canada and South Africa and has been adopted with mixed success in Australia. Its low cost and light weight make it attractive to some home gardeners, particularly where inexpensive, locally available materials are preferred.

One of the main advantages of sawdust culture is its use in free-drainage systems, which can reduce the risk of disease spread. It also provides good lateral movement of nutrient solution through the root zone, along with effective aeration and high water-holding capacity.

However, there are several limitations. Over time, salts can accumulate in the medium — typically from the base upwards — and may reach toxic levels if not managed. This can be controlled through regular leaching with fresh water. As an organic material, sawdust also decomposes during use, leading to a gradual loss of structure. In some cases, this can result in up to a 50 per cent reduction in volume over a growing season, accompanied by reduced aeration.

## Sawdust bag growing system

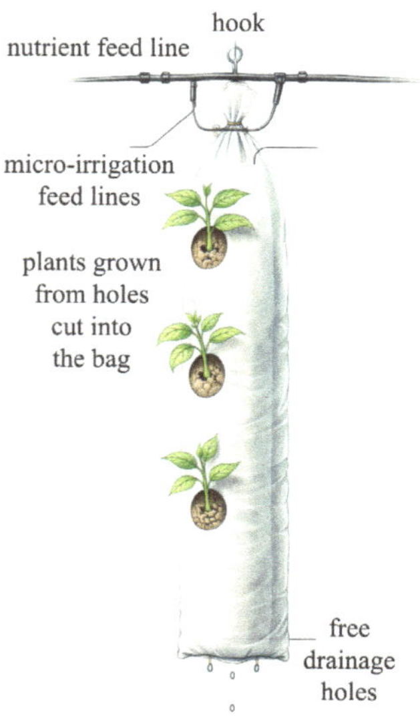


Decomposition is driven by microbial activity, which also increases the demand for nitrogen in the system. As a result, nutrient solutions must be adjusted to compensate for this nitrogen drawdown. Not all sawdust is suitable for hydroponic use, and both hardwood and softwood sawdust — with the exception of jarrah — should be composted or otherwise treated before use to remove potentially harmful compounds.

Sawdust is typically used in either pot or bag systems. Pots are usually 200–250 mm deep with adequate drainage holes, while bag systems use polyethylene bags with perforations to allow free drainage. Bags may be suspended or placed on the ground, although a plastic barrier should be used underneath to prevent roots from growing into the soil. For home gardeners, heavy-duty plastic bags or standard containers such as 20-litre pots can be used effectively.

Drip or trickle irrigation systems are best suited to sawdust culture. Seedlings are often started in rockwool cubes before being transplanted into the sawdust medium.

Because of its ongoing decomposition, sawdust generally needs to be replaced after each crop cycle.

## Aeroponics

Aeroponics is a method of plant cultivation in which roots are suspended in air and intermittently exposed to a fine mist of nutrient solution. The system requires no growing medium, with plants supported above a closed chamber where the root zone is maintained in a humid, oxygen-rich environment.

In operation, the roots are periodically wetted with a nutrient-enriched mist, ensuring that both moisture and essential elements are delivered directly to the root surface. This approach maximises oxygen exposure while maintaining adequate hydration, creating ideal conditions for rapid nutrient uptake.

The effectiveness of aeroponics depends on producing

droplets of the correct size. A fine mist allows for even distribution across the root mass without waterlogging, while also maintaining the delicate balance between air and moisture that is critical to root health. For this reason, nozzle design, pump pressure and timing cycles are key factors in system performance.

*Aeroponic techniques are noted for the development of large root structures.*

Since its development, aeroponics has proven particularly effective for plant propagation, where rapid root formation is desirable. It is widely used in laboratory research into plant physiology and controlled environment agriculture, including studies associated with space-based life-support systems.

The principal advantage of aeroponics is the exceptionally high level of aeration available to the roots. The strong relationship between oxygen availability and root development is well established, with increased air space supporting the intense metabolic activity required for root initiation and subsequent growth.

However, these advantages come with increased technical sensitivity. Aeroponic systems rely on uninterrupted operation; failure of pumps, timers or spray nozzles can quickly lead to root desiccation and plant stress. Regular maintenance and careful monitoring are therefore essential.

While highly efficient in controlled environments, aeroponic systems remain relatively complex and have yet to see widespread adoption at larger commercial scales. Their greatest value continues to lie in specialised applications where precision and rapid plant response are required.

*Diagrammatic cutaway of a triangular aeroponic system showing plant positions and internal mist flow.*

# Choosing a system

With the exception of nutrient film technique (NFT), most hydroponic systems can be adapted to suit a wide range of plants. The choice of system depends on several practical considerations, including the size of your growing area, the type of construction you prefer, the level of automation required and, of course, your budget.

For example, in a small courtyard or balcony where space is limited, a tiered or vertical system can make efficient use of the available area, allowing you to grow a variety of plants for both production and visual appeal. The degree of automation you incorporate will depend largely on cost, as well as your willingness to monitor and maintain the system on a regular basis.

Hydroponic systems are broadly classified as either *open* or *closed*. An open (non-recirculating) system is a free-drainage system in which the nutrient solution passes through the root zone once before being discarded. Common media used in these systems include gravel, scoria, sand, rockwool, coco peat and sawdust.

In closed systems, the nutrient solution is recirculated and reused. These systems include NFT, as well as media-based systems such as rockwool or coco peat culture, flood-and-drain tables, drip-irrigation systems and aeroponic chambers.

## NFT system

The nutrient film technique (NFT) is a closed system that is relatively simple to establish on a small scale. It requires a continuous flow of nutrient solution, which must be adequately aerated to maintain oxygen levels for healthy root development. Aeration can be achieved in several ways, although in practice it is common either to replace the solution at regular intervals or to agitate it using a small aquarium pump and air stone.

While the size, design and construction of NFT systems vary — from large commercial installations to research units and small home systems — the underlying principles remain the same. The essential components include a growing channel to contain the

*Fancy lettuce grown in NFT*

*Some of the different rigid channel profiles available to the home gardener.*

plant roots, a nutrient reservoir, a submersible pump and the necessary irrigation fittings.

The most widely used NFT system for both home gardeners and commercial growers employs UV-stabilised, rectangular, prefabricated rigid channels, fitted with or without removable lids. These channels, often referred to as gullies, are typically mounted on simple, waist-high supports to provide easy access for planting, maintenance and harvesting.

Alternative materials can be used. Rainwater downpipes, for example, can serve as channels, although they are difficult to clean between crops. Round PVC pipe and trays with curved bases have also been used; however, their shape prevents the formation of a consistent nutrient film. In such systems, roots tend to obstruct the flow, leading to flooding and waterlogging. For this reason, these profiles are not recommended despite their lower cost.

A further variation uses inexpensive black-and-white flexible plastic or waterproof tarpaulin. In this arrangement, the channel is formed over a slightly sloping surface to allow the nutrient film to flow by gravity to a collection point for recirculation. Channels are typically about 200 mm wide, with the sides folded and supported to form an inverted 'T' profile. The nutrient solution flows along the base of the channel, with the black inner surface excluding light and the white outer surface reflecting heat.

To assist early plant establishment, a small piece of rockwool or capillary matting can be placed beneath seedlings until their roots extend into the nutrient stream. A support wire positioned about 200 mm above the base helps maintain the channel shape and provides support for young plants.

During operation, it is essential that the 'tent' profile is maintained. The plastic must not be allowed to collapse onto the roots, as this would restrict airflow within the channel and reduce oxygen availability.

*A simple, inexpensive flood-and-drain (ebb-and-flow) system used for raising seedlings. Polystyrene floats support seedlings above the nutrient solution between watering cycles.*

## Flood-and-drain system

The flood-and-drain system (also known as ebb-and-flow or subirrigation) is a closed system widely used by home gardeners, particularly for growing herbs and seedlings. It can utilise a range of growing media, including expanded clay, scoria, river pebbles, rockwool, coco peat and pumice.

The system consists of a shallow growing tray fitted with inlet and outlet controls. At set intervals — typically once an hour, or less frequently depending on environmental conditions — the tray is flooded with nutrient solution for a short period, usually 10 to 15 minutes, before being drained. During the flood cycle, the solution rises to a predetermined level, typically 25 to 50 mm (1 to 2 inches) below the surface of the growing medium. This represents approximately 75 to 80 per cent of the total tray depth, ensuring the medium is thoroughly wetted without submerging the plant stem, after which an overflow allows excess nutrient solution to return to the reservoir for recirculation.

*Terraced flood-and-drain system, with nutrient solution returning to the main reservoir after each drainage cycle.*

This alternating flood-and-drain cycle plays an important role in root health. As the tray fills, stale air within the root zone is displaced upward. As it drains, fresh air is drawn down into the medium. This continuous exchange supplies plant roots with water, nutrients and, critically, oxygen.

Unlike NFT systems, flood-and-drain systems provide a degree of buffering in the event of mechanical failure. The growing medium retains moisture and nutrients, allowing plants to survive for a period if the system stops operating.

The frequency and duration of irrigation cycles are critical to system performance and depend on several factors:

- the type of growing medium
- particle size (particularly with gravel or scoria)
- crop type and stage of growth
- environmental conditions
- time of day

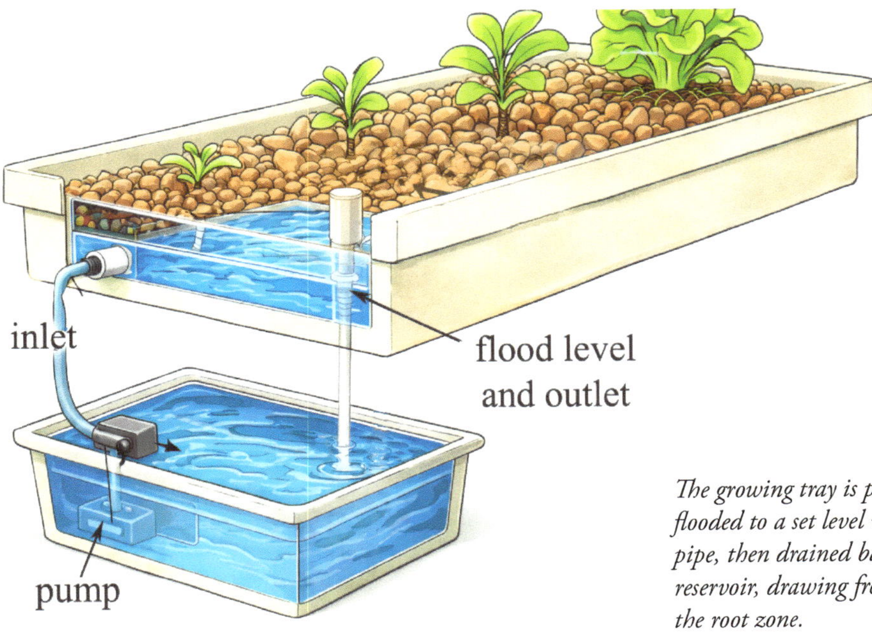


*The growing tray is periodically flooded to a set level via a standpipe, then drained back to the reservoir, drawing fresh air into the root zone.*

Coarse or smooth aggregates require more frequent irrigation than finer, more porous media. Larger, fruiting crops demand more frequent cycles than smaller leafy plants such as lettuce. In hot, dry conditions, or under strong light, evaporation increases and irrigation must be adjusted accordingly.

As a general guide, three to four irrigation cycles per day are suitable in cooler conditions, while in hot summer weather cycles may be required as frequently as once per hour. A small number of cycles during the night is also beneficial. Each cycle should allow the tray to fill and drain rapidly, with complete drainage to prevent residual solution from remaining in the base of the system.

The rate at which the tray floods and drains directly affects root aeration. Roots require oxygen for respiration, which in turn provides the energy needed for water and nutrient uptake. Poor aeration can restrict growth, reduce yields and, in severe cases, lead to plant loss. In practical terms, the greater the movement of nutrient solution, the more effective the exchange of air within the root zone.

## Drip-irrigation system

Drip irrigation (also known as trickle irrigation) can be operated as either an open or closed system, although most home gardeners prefer the open, free-drainage option. These systems are relatively simple to design and construct, using components that are readily available from local hardware or irrigation suppliers.

A key requirement of drip systems is that the nutrient solution must be completely soluble in water. Any undissolved material can clog micro-irrigation tubing and emitters, leading to uneven distribution of nutrients. For this reason, liquid concentrates are generally preferred, and care should be taken to ensure that no sediment remains in the solution. The use of an in-line filter in the main feed line will further reduce the risk of blockage.

*Coriander grown in a bench-height sand system using drip irrigation lines.*

Drip systems can be used with a wide range of growing media, including rockwool, coco peat, expanded clay, gravel, sand, perlite/vermiculite, scoria, pumice, sawdust and zeolite. The watering regime depends on the characteristics of the chosen medium, as well as environmental conditions such as temperature and light.

One of the main advantages of drip irrigation is its efficiency. Nutrient solution can be delivered directly to the root zone in controlled amounts, reducing waste and allowing the grower to tailor irrigation frequency and duration to the needs of the crop at different stages of growth. Because the nutrient solution is applied gradually, good aeration is maintained within the root zone, and plant roots are never submerged.

However, drip systems are not without their limitations. In addition to the risk of clogging, coarse media such as gravel can lead to uneven water distribution. This may result in 'coning', where the nutrient solution travels vertically downward rather than spreading laterally through the root zone. In such cases, plants may experience water stress, while roots tend to concentrate at the base of the bed, potentially blocking drainage outlets.

The layout of a drip-irrigation system is straightforward. A 13 mm or 10 mm poly supply line runs from the nutrient reservoir along the length of the growing bed and is sealed at the far end. From this main line, 4 mm micro-tubing (commonly referred to as spaghetti tubing) branches off at regular intervals to deliver nutrient solution directly to the base of each plant. Flow can be regulated using adjustable emitters fitted to the micro-tubing, or by installing a simple tap or valve in the main supply line near the reservoir. This allows the overall pressure and delivery rate to be controlled. In some configurations, excess nutrient solution can be returned to the reservoir, where it helps

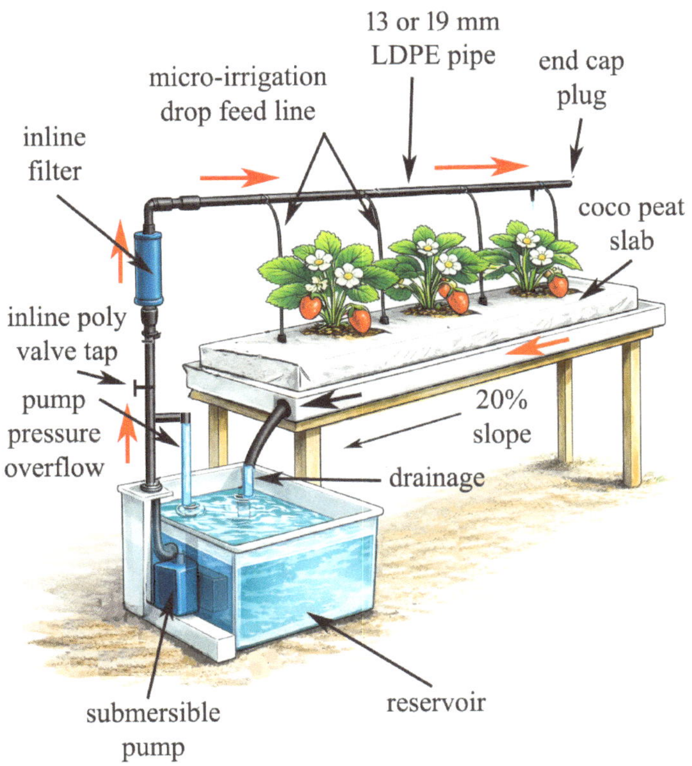


*Recirculating drip irrigation system. Drainage from the slab and pump pressure overflow both return above the reservoir, promoting aeration of the nutrient solution.*

to aerate the solution before recirculation.

Beds should be well drained, with excess nutrient solution either allowed to run-to-waste or collected and returned to the reservoir for reuse.

## Aeroponic system

In an aeroponic system, plant roots are suspended within an enclosed chamber and intermittently wetted by a fine mist of nutrient solution. Unlike other hydroponic techniques, no solid growing medium is used. The roots are maintained at close to 100 per cent relative humidity to prevent desiccation, while the chamber itself must be both moisture-proof and light-excluding to inhibit algal growth and maintain a stable root environment.

The design of aeroponic systems can vary widely. Chambers may be constructed in vertical or horizontal configurations depending on the crop and available space. Tomatoes are often grown in tall, narrow containers lined with plastic, while lettuce and strawberries are well suited to A-frame arrangements that maximise both light exposure and space efficiency.

The principal advantage of aeroponics lies in the exceptional level of root aeration it provides. Because the roots are suspended in air and exposed to a fine nutrient mist, oxygen is readily available to support the intensive metabolic processes associated with root development and plant growth. Under ideal conditions, this can lead to rapid growth rates and efficient nutrient uptake.

*Aeroponic research system cultivating green oak lettuce, Environmental Research Laboratory, University of Arizona, Tucson.*

However, these advantages are offset by significant practical limitations. Aeroponic systems are mechanically complex and require careful regulation of water delivery, nutrient concentration and environmental conditions. Unlike systems that incorporate a growing medium or a reservoir of solution, aeroponics has little or no buffering capacity. As a result, even minor interruptions — such as blocked nozzles, pump failure or power loss — can lead to rapid plant stress and, in some cases, irreversible damage within a short period.

Despite these limitations, aeroponics has proved highly effective for propagation. Experimental work has demonstrated a strong relationship between dissolved oxygen levels and root development. Difficult-to-root species such as chrysanthemum and ficus have shown improved root initiation, increased root numbers and greater overall root length when exposed to higher oxygen concentrations. The technique has also been applied successfully on a commercial scale, including large propagation programmes to produce millions of saplings for reforestation, significantly reducing production time.

More sophisticated adaptations of the technique have sought to overcome its inherent limitations. The Ein Gedi System (EGS), developed in Israel, combines elements of aeroponics and water culture. In this system, plant roots are exposed simultaneously to a continuously aerated nutrient solution and a fine spray delivered above it. Oxygen is transferred from the spray zone into the liquid phase, maintaining high oxygen concentrations at the root surface while providing a larger reservoir of solution. This hybrid approach allows roots to develop within both the spray zone and the liquid environment, improving system stability while retaining the benefits of enhanced aeration.

Other variations include systems such as the Rainforest unit, in which a submersible pump delivers nutrient solution upward to a deflector plate, creating a circular spray pattern that wets the roots before returning to the reservoir. Experimental systems using ultrasonic technology have also been developed to produce an extremely fine mist; however, issues such as overheating and equipment reliability have limited their practical application.

In theoretical terms, aeroponics represents one of the most

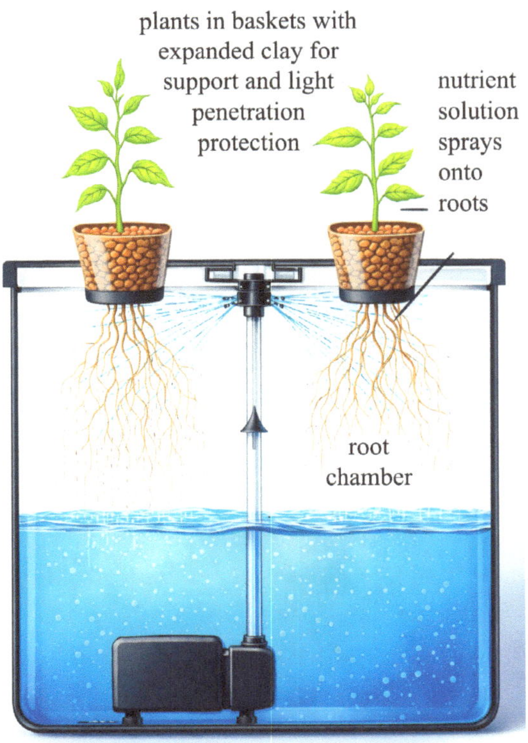


*A system where the roots of a plant are continuously or intermittently in an environment saturated with fine drops (a mist or aerosol) of nutrient solution.*

efficient methods of delivering water, nutrients and oxygen to plant roots. In practice, however, its adoption is constrained by cost, complexity and the need for precise control systems. For most home gardeners, these requirements place aeroponics beyond practical reach. Nevertheless, it remains an important technique for plant propagation, research applications and specialised commercial production where its advantages can be fully realised.

*Built-in pH monitoring system within an aquaponic unit, allowing continuous tracking and timely adjustment of water chemistry.*

## Aquaponics

Aquaponics is the integration of recirculating aquaculture (fish culture) with hydroponic plant production. In this system, water from a fish tank—enriched with fish waste—is continuously circulated through plant growing areas, where it serves as a nutrient source. Naturally occurring bacteria convert the waste products into forms readily absorbed by plants, and in doing so, effectively purify the water before it is returned to the fish. The result is a closed-loop system in which fish, plants and bacteria operate in a stable, interdependent cycle.

*Aquaponic system occupying 2.5 square metres, with the fish tank located beneath the elevated hydroponic unit. Total system volume is approximately 850–900 litres, supported by an 80 litres per minute aeration pump.*

Unlike conventional hydroponics, where nutrient solutions are formulated and adjusted manually, aquaponics relies on biological processes to generate nutrients. Fish waste provides ammonia, which is converted by nitrifying bacteria first into nitrite and then into nitrate—the primary nitrogen source for plant growth. This biological filtration is central to system stability and must be established before the system reaches full productivity.

When choosing an aquaponic system, the grower must consider three essential components: the fish rearing unit, the filtration system (both solids removal and biological filtration), and the plant growing system. These components must be balanced so that nutrient production from the fish matches plant uptake, while maintaining water quality suitable for aquatic life.

*Backyard aquaponic system capable of supporting up to 15 fish and more than 20 vegetable plants.*

Several system configurations are commonly used. For small-scale or backyard applications, media-filled beds (typically gravel or similar aggregate) are widely adopted. These systems are simple and versatile, as the growing medium can also function as both a biological filter and, to some extent, a solids trap. However, care must be taken to prevent clogging and the formation of anaerobic zones, which can adversely affect both plant and fish health.

For larger or more intensive systems, media-free hydroponic techniques are often preferred. Deep flow systems (also known as raft systems) allow plants to grow on floating boards with roots suspended directly in the nutrient-rich water. These systems are relatively easy to manage and well suited to leafy crops. Nutrient Film Technique (NFT) systems may also be used, in which a shallow stream of water flows through channels, providing nutrients while maintaining a high level of root aeration.

In selecting a system, consideration must also be given to local conditions, particularly the choice of fish species. Different species have varying tolerances to temperature, oxygen levels and water quality, and these factors will influence system design, filtration requirements and overall management.

*Aquaponic systems range from simple backyard media beds and compact indoor units to raft and flood-and-drain designs, as shown here with a fast-growing water spinach (Kangkung) crop.*

Aquaponics offers several advantages, notably efficient water use, minimal waste discharge and the ability to produce both plant and protein crops from a single input. However, it also introduces an additional layer of complexity compared to standard hydroponics, as the needs of both plants and fish must be met simultaneously. Successful operation therefore depends on maintaining a careful balance between biological processes, water quality and system design.

A reliable system depends on consistent water circulation and adequate aeration to support fish and bacteria. Temperature must suit both fish and plants. Regular monitoring of pH and nitrogen levels is essential. Stocking density and feeding must be controlled, as excess inputs can quickly overwhelm the system's biological capacity.

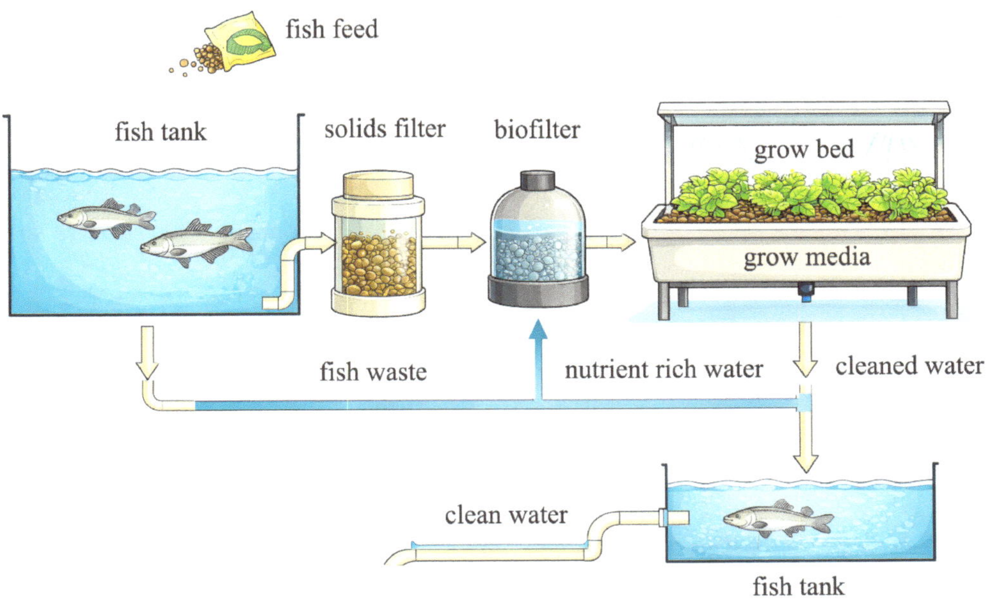

*Pot-based wick system, with nutrient solution drawn from the reservoir to the root zone by capillary action.*

# Other systems

## Wick system

The wick system is a simple closed system and remains one of the most widely used hydroponic methods for home gardeners. It consists of a double container arrangement, with one section holding the growing medium and plant, and the other containing the nutrient solution. A fibrous wick—commonly hemp, cotton or synthetic fibre—is inserted into the growing medium, typically about one-third of the way up, with the opposite end suspended in the nutrient reservoir below.

As moisture is taken up by the plant and lost through transpiration, capillary action draws nutrient solution from the reservoir through the wick and into the root zone. This passive movement of solution ensures a continuous, though relatively slow, supply of water and nutrients without the need for pumps or mechanical devices.

The system is particularly suited to small plants and low-demand crops, as the rate of nutrient delivery is limited by the properties of the wick and growing medium. For this reason, free-draining media such as perlite, vermiculite or coco fibre are preferred to maintain adequate aeration around the roots.

Wick systems can be easily adapted for use on balconies, window ledges or in under-utilised areas of the garden, and may be supplied from a single central reservoir servicing multiple containers. Although not essential, the inclusion of a simple float or level indicator is useful in monitoring nutrient solution levels and maintaining consistent operation.

*Gravity-feed hydroponic system, with nutrient solution delivered from an elevated reservoir and collected below for reuse.*

## Gravity-feed system

The gravity-feed system, which may be operated as either an open or closed system, is popular with home gardeners due to its low cost and simplicity of construction and operation. Its effectiveness depends on the correct positioning of components, with the nutrient reservoir located above the growing bed and a collection reservoir positioned below.

Nutrient solution flows by gravity from the upper reservoir into the growing medium, where it is distributed through the root zone before draining into the lower collection receptacle. From there, the solution is returned to the nutrient reservoir using either a foot pump or a small submersible pump.

In its simplest form, the system can be operated without

electricity by periodically interchanging the nutrient and collection containers, making it an efficient and practical option for low-input hydroponic cultivation.

*Three small Leni pots with water gauge, planted with ornamentals in expanded clay, demonstrating a decorative, individual container approach to hydroponic growing.*

## Gravity/vacuum system

The gravity/vacuum system was developed in the early 1990s, applying a well-established irrigation principle in a simple closed configuration. It consists of a growing tray coupled to an air-tight nutrient reservoir.

The water inlet to the growing tray is set at the same level as an air line returning to the reservoir. Gravity initially allows nutrient solution to flow into the tray, forming a shallow pool at the base of the growing bed. Once the solution level reaches the height of the air line, a partial vacuum is created within the sealed reservoir, effectively halting further inflow and stabilising the nutrient level within the tray.

The system operates without pumps, timers or mechanical controls, relying on the interaction between gravity and vacuum to regulate nutrient flow. This simplicity makes it highly reliable, with minimal maintenance and low operating costs. As solution is taken up by the plants or lost through evaporation, the vacuum is released and replenishment occurs automatically, maintaining a consistent nutrient level within the growing tray.

Most components required for construction are readily available. UV-stabilised growing trays, 20-litre plastic reservoirs, irrigation hose and standard fittings can be sourced from hardware or irrigation suppliers. A key element in maintaining the vacuum seal is a correctly fitted rubber O-ring at the reservoir connection, ensuring the system remains air-tight and functions as intended.

*Gravity/vacuum system illustrating a simple, self-regulating design based on the principle used to maintain constant water levels in chicken shed drinkers.*

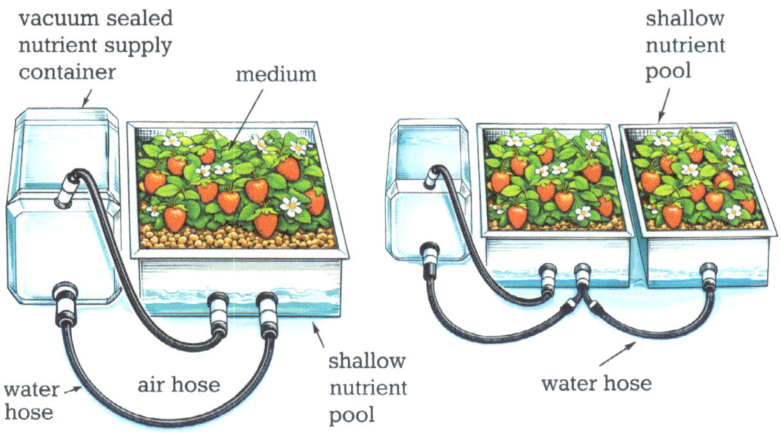

## Self-watering pots

Self-watering pots are a simple form of closed hydroponic system and provide an effective entry point for beginners. A number of commercial designs are available, typically consisting of a combined pot and reservoir arrangement that maintains a constant supply of nutrient solution to the plant.

Common designs incorporate either an attached reservoir at the base of the pot or a separate outer container holding the nutrient solution. Many systems include a simple water level indicator, allowing the grower to monitor solution levels and maintain consistent operation.

Expanded clay is the most widely used growing medium in these systems, valued for its attractive appearance, durability and reliable capillary action. The medium provides both support and aeration while allowing moisture to move freely to the root zone.

Most self-watering pots use a double-container design, with a perforated inner pot seated within a sealed outer reservoir. The perforations allow roots to access both air and nutrient solution, promoting healthy growth under low-maintenance conditions. As plants develop, the inner pot can be transferred to a larger container and surrounded with additional medium, allowing for easy upscaling.

These systems are particularly well suited to indoor environments, including homes, offices and public spaces, where they offer a clean, reliable and low-maintenance method of plant cultivation.

*Pot-based wick system, with nutrient solution drawn from the reservoir to the root zone by capillary action.*

*Side-porch balcony garden growing tomato and bell pepper in plastic pots and buckets, using perlite as the growing medium.*

*Round downpipe with cutaways adapted as a sand and gravel system for growing strawberries, harvested every second day.*

*Strawberries grown in a home-made NFT system constructed from PVC stormwater guttering, propagated in rockwool cubes, then transferred to the NFT system.*

*Brussel sprouts growing in a reciculating aggregate system. The main substrate is blue metal which provides good drainage and aeration. These plants are fed via a trickle irrigation system.*

*Chinese turnips grown in a sand and blue metal aggregate mix, irrigated by drip feed for 15 minutes at two-hour intervals. The system is set at waist height for ease of access.*

*Drip-irrigated aggregate system using a 40% nursery sand and 60% blue metal mix, supplied from a central nutrient reservoir via 19 mm poly tubing.*

*Stacked container hydroponic system using perlite as the growing medium, arranged for gravity-fed drip irrigation with likely drain-to-waste nutrient management.*

*Backyard gravity-fed (vacuum-assisted) system growing corn, leafy greens and cherry tomatoes, all sharing a compatible nutrient regime.*

*Cos (Romaine) lettuce grown in a closed-loop NFT system, with a continuous thin film of nutrient solution flowing over the roots in channels.*

*Rocket (arugula) grown in a space-saving vertical NFT system, a fast-growing cool-season leafy green with a distinctive peppery flavour.*

*Capsicum (peppers) grown in a rockwool hydroponic system, providing excellent aeration and moisture control for healthy root development.*

*Spring onions harvested from a small polystyrene box filled with a perlite and vermiculite mix, supplied by trickle irrigation.*

# Building a system

The simplest and most economical way to begin hydroponic gardening is through the use of plastic or polystyrene trays and containers. These are widely available and can often be sourced at little or no cost from local fruit and vegetable outlets. Polystyrene produce boxes, particularly those used for broccoli, are well suited to adaptation as hydroponic growing beds, as they typically have solid bases without drainage slits.

To convert these containers into functional hydroponic units, drainage holes should be made approximately 20 mm above the base. This allows a shallow reservoir of nutrient solution to form within the tray, from which plants can draw moisture and nutrients as required. The principle is similar to that used in self-watering pots, where a small standing volume of solution provides a consistent and accessible supply to the root zone.

Irrigation can be carried out manually or automated through the use of simple micro-irrigation systems. These components—including small-diameter tubing, emitters and connectors—are readily available from nurseries, garden centres and specialist hydroponic suppliers. A basic nutrient reservoir may consist of a plastic pail or drum, with a small submersible pump, such as those used in aquariums, used to distribute the solution to the growing trays.

For a free-drainage system, polystyrene boxes with existing drainage slits may be used without modification. These are typically filled with a suitable aggregate medium such as a sand-based or perlite vermiculite mix, and supplied with nutrient solution via a trickle irrigation system. Care must be taken to regulate irrigation cycles so that only a small proportion of the applied solution drains from the system. As a general guide, excess drainage should not exceed approximately ten per cent of the total volume applied.

*Assembling a small-scale hydroponic growing system.*

This can be monitored by placing a small container beneath one of the trays to collect runoff during irrigation. After each cycle, the collected volume provides a simple indication of system efficiency. Achieving the correct balance may require a period of adjustment, but once established, the system will operate reliably with minimal intervention.

These simple systems provide an ideal introduction to hydroponic cultivation, allowing the grower to develop practical skills and confidence before progressing to more advanced designs. As experience is gained, systems can be expanded, refined or replaced with more sophisticated arrangements to suit individual requirements and available space.

## Nutrient Film Technique (NFT) Systems

A wide range of small-scale NFT systems is available commercially, and there are sound reasons for choosing a ready-made unit. The most obvious is convenience: a complete kit that can be assembled quickly and put into operation with minimal effort. For those with limited time, or for gardeners seeking a straightforward introduction to hydroponics, these systems offer a practical and reliable starting point.

However, an NFT system can just as effectively be built to suit individual requirements. Custom construction allows flexibility in size, layout and materials, enabling the system to be tailored to the available space and the crops being grown.

While designs may vary, all NFT systems must satisfy a small number of essential requirements. The system must provide a means to:

- Support the plant above the nutrient stream
- Deliver nutrient solution in a controlled, continuous flow
- Maintain adequate aeration within the solution
- Exclude light from the root zone to prevent algal growth

These principles form the foundation of every successful NFT installation, regardless of scale.

Constructing a simple NFT system requires only basic skills and readily available materials, and for many home gardeners it becomes an enjoyable and rewarding project. In practice, system size is an important consideration. Most home installations fall within a range of two to five metres in length, with three to five growing channels.

Excessively long channels are best avoided. As channel length increases, temperature variations can develop along the nutrient flow, circulation may slow, and plants positioned toward the end of the system can receive reduced aeration.

*A variety of home hydroponic systems are available, ranging from commercially manufactured units to simple designs constructed from readily available materials.*

*Custom-built 3-metre, four-channel recirculating NFT system, incorporating a trellis and heavy-gauge netting to support tomato plants during growth.*

*The use of round pipes, such as standard PVC stormwater pipe, is common in small-scale, do-it-yourself systems.*

*Deep flow technique (DFT) system using large-diameter PVC pipes, where plants are supported above a relatively deep, recirculating nutrient solution.*

*Home-built horizontal NFT system supporting four growing channels, simple to construct and operate.*

*Vertical recirculating NFT system constructed from PVC plumbing pipe and fittings, demonstrating a compact, space-efficient growing arrangement for leafy crops.*

Maintaining shorter runs helps ensure more uniform growing conditions across all plants.

Site selection is equally important. NFT systems should be positioned in a location protected from strong winds, which can increase evaporation and disrupt plant stability. Where a sheltered site is not available, the use of a simple windbreak can significantly improve system performance.

## Channel layout

The standard NFT system uses rigid channels, although simpler installations can be constructed using trays. In multiple-row layouts, channel spacing should generally be between 10 and 30 cm, depending on the crop being grown.

Channel size is an important consideration. Configurations range from approximately 50 mm wide by 40 mm high to 300 mm wide by 70 mm high. For leafy vegetables, a commonly used size is 150 mm wide by 70 mm high. The dimensions selected will largely determine the type of crop that can be grown successfully. Standard channels are well suited to lettuce, endive, strawberries and many Mediterranean and Asian vegetables and herbs. Larger plants such as tomatoes and cucumbers are less suitable, as their extensive root systems can restrict flow, leading to blockage and 'ponding', where nutrient solution accumulates rather than moving freely.

Channels or trays should be installed with a gentle slope, typically between 1:30 and 1:50, to ensure continuous flow and prevent waterlogging. Even with correct fall, some ponding may occur as root systems develop. The required slope can be achieved through the floor, benches, support frames or the natural lay of the land.

*Hydroponic strawberries grown in a DFT system. Expanded clay aggregate provides plant support above a relatively deep, recirculating nutrient solution.*

As a guide, flow rates for each channel should be around 1 litre per minute. While this may vary between installations, rates should generally not fall below 0.5 litres per minute or exceed 2 litres per minute, as extremes can lead to nutritional imbalance or poor plant performance.

There is no fixed irrigation regime for NFT systems. Most operate with a continuous flow of nutrient solution, although some growers adopt intermittent cycles, typically a few minutes of flow every 10 to 20 minutes. Intermittent operation can improve root aeration, but in warm conditions there is a fine balance between adequate aeration and the risk of roots drying out.

The use of round pipes, such as standard PVC stormwater pipe, is common in small-scale and do-it-yourself systems, but introduces several limitations when compared with flat-bottomed channels. The principal issue is that the curved base makes it difficult to maintain a thin, even film of nutrient solution—the defining feature of an efficient NFT system. Instead, solution tends to collect along the lowest point of the pipe, increasing the likelihood of uneven flow and localised ponding.

The circular profile also encourages roots to accumulate in a narrow, deeper mass, which can further restrict flow as plants mature. This not only affects nutrient delivery but also complicates cleaning between crops. In addition, it is often difficult to achieve a complete light seal around planting holes or net pots, allowing light to enter the pipe and promote algal growth.

While round pipes remain an inexpensive and readily available option, they generally result in higher maintenance requirements and less consistent performance than purpose-designed flat channels.

## Channel slope

The NFT channel must be set with a firm, uniform slope to prevent 'ponding', where roots lose contact with air and become waterlogged. A minimum slope of 1:50 is generally accepted, although a slightly steeper fall is preferable where practical. In some cases, achieving a greater slope may be limited by the additional lift required from the pump.There is no strict upper limit to channel slope, and successful crops have been grown in vertical NFT systems. The main constraint in such arrangements is the ability to securely support plants within the channel.

*Compact home NFT system (1.2 m long, four channels wide) growing Cos (Romaine) lettuce.*

## Channel width

For healthy vine crops such as cucumbers and tomatoes, narrow channels are unsuitable. These plants develop extensive root systems, and this is no different in hydroponic culture.To accommodate the root mass, channels should be at least 250 mm wide at normal plant spacing. Wider channels can be constructed at ground level using heavy black and white plastic to form simple, effective growing beds.

## Reservoir volume

The volume of working nutrient solution in the reservoir is critical to the efficient operation of an NFT system and will vary according to the crop being grown. As a general guide, allow approximately 1 litre of nutrient solution per plant. For example, a system supporting 20 plants should have a reservoir capacity of about 20 litres.

If the volume is too low, plants may take up water more rapidly than nutrients, causing a rise in nutrient concentration. Conversely, excessive volume can dilute nutrient levels, reducing availability to the plants. In either case, rapid fluctuations in concentration can occur over a short period.

Elevated nutrient levels may lead to root damage, while low concentrations will restrict growth. These changes are often accompanied by shifts in pH, which can place additional stress on plants and result in a range of growth disorders. Maintaining an appropriate reservoir volume therefore helps stabilise both nutrient concentration and pH, ensuring consistent plant performance.

*Wet hessian bag used to insulate the reservoir and moderate nutrient solution temperature in hot conditions.*

## Reservoir site

The nutrient reservoir determines the direction of flow within the channels and should be positioned as the lowest component of the NFT system. It must also be covered to prevent rainwater from entering and diluting the nutrient solution. In extreme hot weather, covering the exposed reservoir with a wet hessian bag can help moderate and stabilise the temperature of the nutrient solution.

## Solution temperature

The ideal temperature of the nutrient solution varies between crops, but for most plants it should be maintained between 18°C and 25°C. There is a risk of overheating if channels are exposed to prolonged direct sunlight during the warmer months.

As plants mature, foliage will provide some shading to the channels, reducing the likelihood of excessive temperature rise.

## Garden reflectors

Most vegetables require at least 6–8 hours of direct sunlight each day for satisfactory growth. In small urban gardens, this can be difficult to achieve where growing areas are confined between fences, walls or buildings, resulting in prolonged shading.Where light is limited, simple reflector panels can be used to increase the amount of available light. These can be made by fixing aluminium foil or similar reflective material to sheets of cardboard or plywood. Panels should be positioned to reflect sunlight back onto the growing area and, where possible, made to match the length of the garden.If a suitable wall is available, painting it white will improve light reflection and may also provide a small amount of retained heat, which can be beneficial during cooler periods.

# DIY Mobile Vertical NFT Hydroponic System

The following example outlines the construction of a compact, mobile vertical NFT system suitable for home use. The design is modular, allowing components to be adjusted to suit available space and materials. While many variations are possible, this design demonstrates the basic principles common to most small-scale NFT systems. *(Design Peter Doyle)*

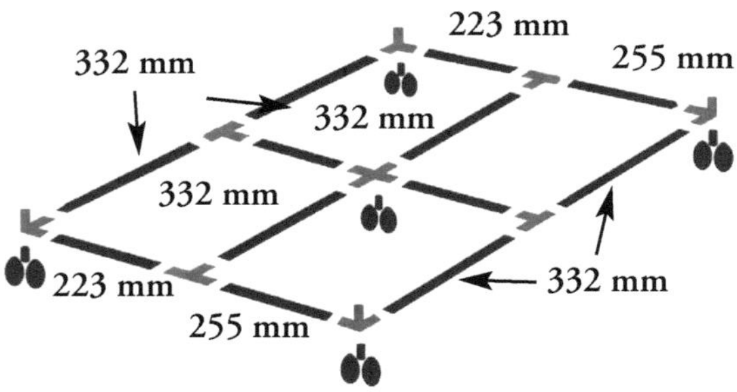


**Step 1: Base frame assembly**

4 x 3-way corners
1 x 4-way cross
4 x 3-wat Tee
6 x 322 mm square tube
3 x 255 mm square tube
3 x 232 mm square tube
5 x castor wheels

**Step 2: Upper frame**

4 x 3-way corners
1 x 3-way Tee
2 x 223 mm square tube (* HDC attached)
2 x 255 mm square tube (* HDC attached)
4 x 400 mm square tube
2 x 700 mm square tube
* *HDC (Hold Down Clip)*

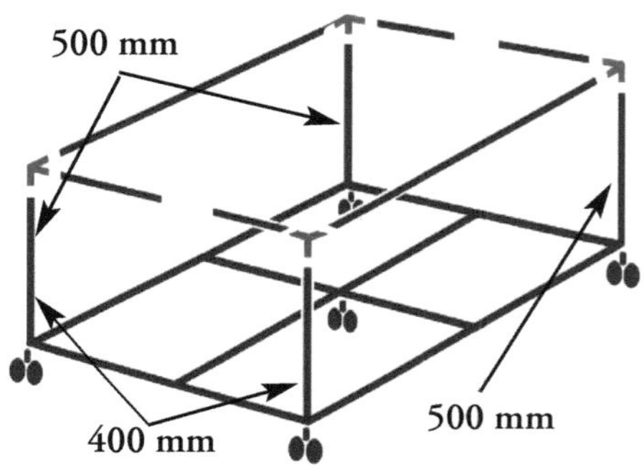


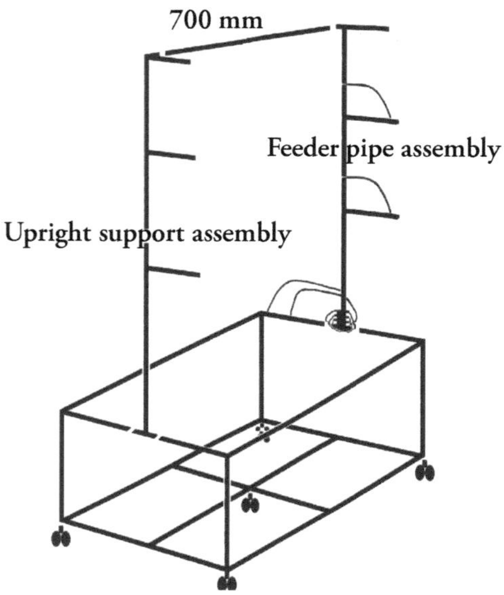


**Step 3: Upright supports**

1 x Feeder pipe assembly
1 x Upright support assembly
1 x 700 mm square tube

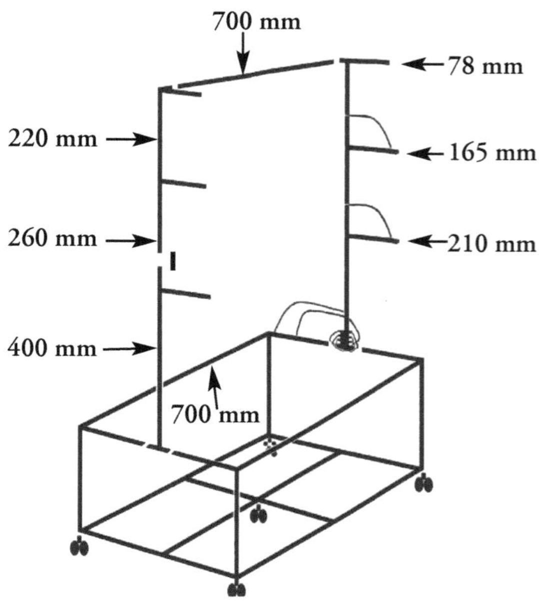


**Step 4: Channel supports**

6 x channel support arms

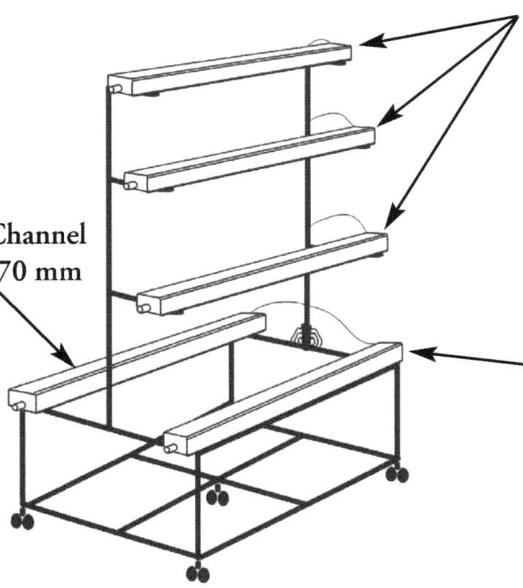


**Step 5: Channel placement**

Place channel on supports
Hold down with clips

5 x 100 mm end caps
5 x 100 mm end caps with spout

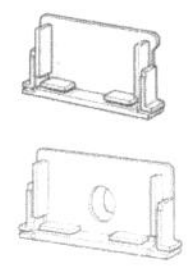

**Step 6: Return manifold**

3 x pieces

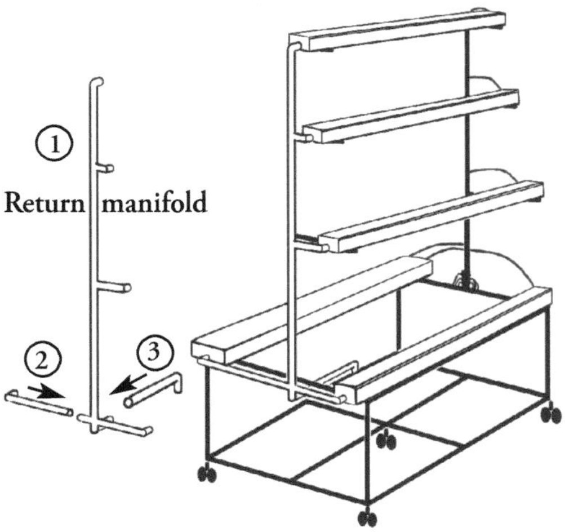


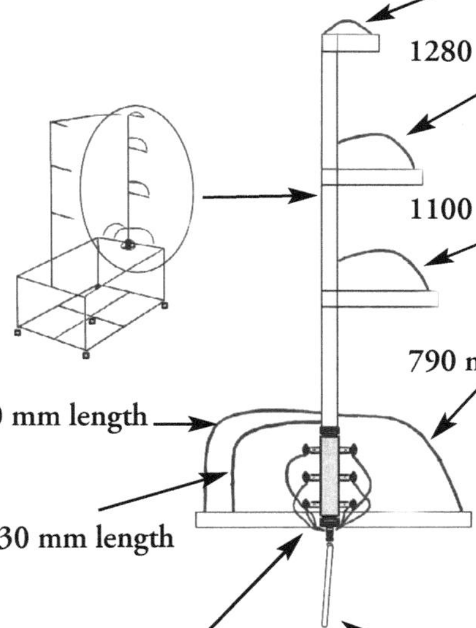


**Step 7: Feeder assembly**

4 mm white poly (spaghetti) tube

Once assembled, the system should be tested with water before adding nutrient solution. Check all joints for leaks, confirm even flow through each channel, and adjust the slope if necessary to ensure a continuous, shallow film.

The NFT system described above represents one of the most efficient and widely used hydroponic techniques. The following sections outline alternative systems, each suited to different crops, environments and levels of experience.

# Flood & Drain System (Closed Loop)

Flood-and-drain systems are among the most widely used methods in hydroponics, valued for their simplicity, reliability and versatility. The principle is straightforward: the growing tray is periodically flooded with nutrient solution and then allowed to drain, drawing fresh air down through the root zone as the solution recedes.

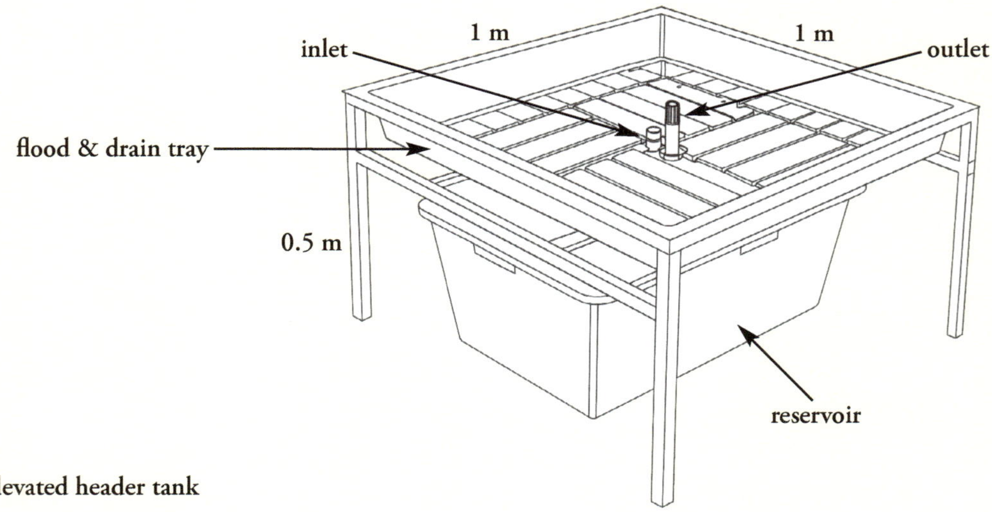


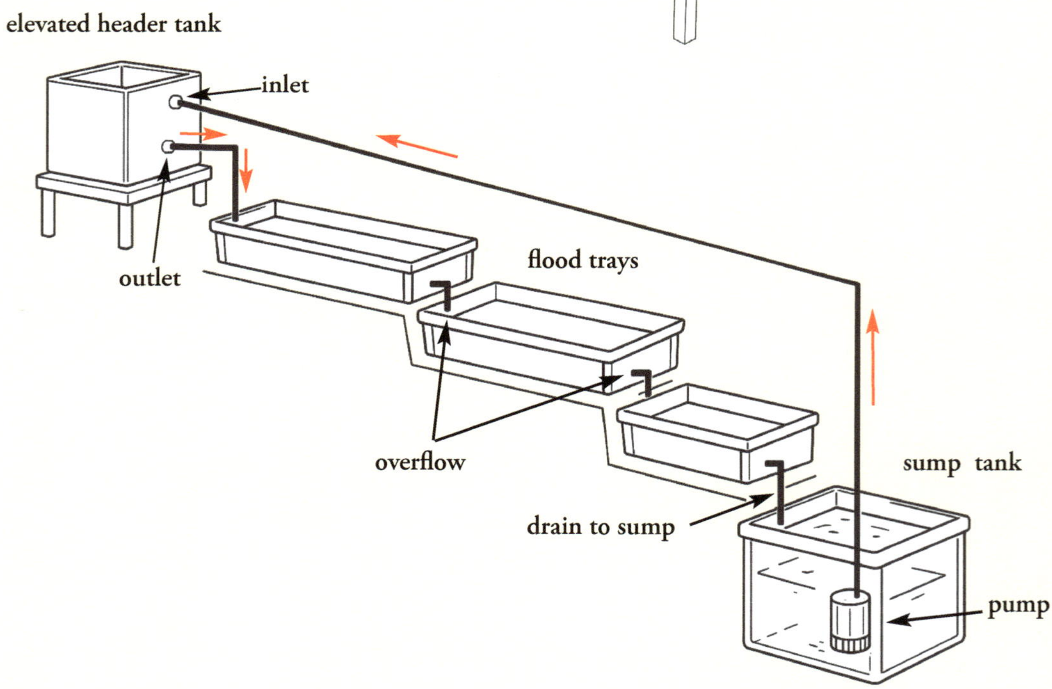


In this terraced arrangement, **gravity does most of the work**. Solution flows from a header tank through each tray in turn before returning to the reservoir, where it is recirculated. The system is efficient, easy to construct, and well suited to a wide range of crops and growing media.

## Flood-and-drain system

Flood-and-drain systems are among the easiest hydroponic systems to build and operate, making them ideal for home gardeners. You can construct them as tables, trays, or simple bucket units, and expand the system as your needs grow.

When building the system, ensure it is capable of rapid filling and complete drainage. Use inlet and outlet lines of at least 19 mm diameter; alternatively, make the outlet larger than the inlet to prevent slow drainage and waterlogging.

Install a timer to control the pump and regulate the irrigation cycle. This allows the growing medium to flood-and-drain at regular intervals, supplying both nutrients and oxygen to the roots. Use opaque tanks and delivery lines to reduce light penetration and minimise algae growth.

## Flood-and-drain tables

Flood-and-drain tables can be arranged as single units or as multiple trays. These may be set at the same height or stepped in a terraced layout. A terraced arrangement is particularly useful on uneven ground and allows the system to operate with partial gravity feed.

When constructing terraced systems, position each lower bed slightly below the one above it. As a guide, the base of the upper bed should align with the top of the next bed.

Build growing beds to a minimum depth of 12 cm to ensure adequate root development and moisture retention. Fit an overflow valve and set it so the nutrient solution rises to within approximately 10 mm of the top of the growing medium before draining. In single-table systems, the solution returns directly to the reservoir; in terraced systems, it flows from one tray to the next.

## Nutrient reservoir

Position the nutrient reservoir according to the system layout. For single or level tables, place the reservoir beneath the first table and use a submersible pump to deliver the solution.

For terraced systems, install an elevated header tank at least one metre above the highest growing bed. Control the flow using either an automatic solenoid valve or a manual gate valve.

Allow the nutrient solution to move through the system by gravity, draining into a sump at the lowest point. From there, pump the solution back to the header tank to complete the cycle. Regularly check flow rates and return lines to ensure efficient circulation and prevent blockages.

*Raised media-bed flood and drain (ebb and flow) hydroponic system filled with expanded clay, illustrating a simple, scalable design suitable for backyard construction.*

**Building tips**

• Building a hydroponic system, like any practical project, is easier and more enjoyable with the right tools and equipment. Think in terms of basic plumbing.

• When selecting a system, consider the time required for maintenance. A small recirculating unit capable of supporting up to 50 plants may require as little as 10 minutes per day to check pH, monitor nutrient levels, and carry out routine adjustments.

• Plant only what you are likely to use during the harvest period. Plan ahead—visualise the yield and match planting to your needs. If unsure of performance, allow a small margin for losses, but avoid planting more than twice your intended harvest.

## Growing medium (substrate)

The choice of growing medium is fundamental to the success of a flood-and-drain system. The medium must provide firm support for the plant while allowing roots access to both water and oxygen.

When selecting a medium, consider its ability to retain moisture without becoming waterlogged. Good aeration is essential, as plant roots require oxygen to function properly. Equally important is effective drainage, ensuring the nutrient solution can move freely through the bed during each flood-and-drain cycle.

Practical considerations also play a role. Readily available materials such as washed river sand or gravel are often suitable, provided they are clean, inert, and free of fine particles that may restrict drainage. Cost and ease of sourcing should be weighed against performance.

Lightweight substrates such as perlite may be used in blends, but on their own they are often too light to adequately anchor larger plants. Expanded clay aggregates (often sold as clay balls or LECA) are a popular alternative among home growers. Although more expensive, they are clean, reusable, and provide an excellent balance of aeration and drainage.

In all cases, the medium should remain stable over time and not break down or compact, which would reduce both aeration and drainage.

While flood-and-drain systems suit many crops, drip irrigation offers greater control over nutrient delivery and is widely used in both small and large-scale systems.

*Drip-irrigation system with an overhead main feed line and 4 mm micro-adaptors delivering nutrient solution to the growing medium and plant.*

## Drip-irrigation system

A simple drip-irrigation system can be constructed using a growing tray or bed, a nutrient reservoir, and a trickle-feed delivery system operated by a pump on a timer. The system is adaptable, relatively low-cost, and can be expanded as required. It is particularly suited to a wide range of crops, from leafy greens to fruiting plants, where controlled and consistent moisture delivery is important.

A wide range of growing media may be used, including rockwool, coco peat, gravel, sand, perlite–vermiculite mixes, expanded clay, or sawdust. Each has its advantages, with the final choice usually determined by cost, availability, water retention, and aeration characteristics. In practice, many growers experiment with blends to achieve the best balance between moisture retention and drainage.

## Growing beds

Growing beds may be constructed at ground level or raised, although ground-level beds are generally less labour-intensive. A typical bed is about 1 metre wide and 400 mm deep, with a slight fall of approximately 1:400 to promote effective drainage and prevent pooling of nutrient solution.

The bed should be filled unevenly—within about 20 mm of the top at the inlet end and 40 mm at the outlet—to ensure even distribution of nutrient solution along its length. Filling the bed level may result in uneven wetting of plant roots and inconsistent growth.

Beds can be formed using timber planks or concrete reinforcement mesh shaped to the required dimensions. Line the bed with a durable plastic membrane, such as pond liner, to prevent nutrient loss into the soil. If using polyethylene film, double it over to improve strength and reduce the risk of tearing under load.

Install a perforated 90 mm drain pipe (agpipe) along the full length of the bed, connected to a main outlet pipe to remove excess solution. When using sand, cover the pipe with coarse gravel (blue metal) to prevent clogging and discourage root intrusion. This also helps maintain consistent drainage over time.

Trickle-fed systems can also be readily adapted to containers such as pots, buckets, and small trays. These simple set-ups are ideal for home gardeners and can often be assembled using materials already at hand. While smaller in scale, they operate on the same principles of controlled delivery and drainage, making them an effective and economical alternative to full-sized growing beds.

*Young cucumbers established in individual buckets using a perlite–rockwool medium, illustrating a simple modular hydroponic system suited to small-scale production.*

## Nutrient reservoir

The nutrient reservoir must be watertight and sized to hold 30–40% more than the volume required to fill the growing bed. This ensures a stable supply during operation and allows for minor fluctuations in usage.

Fit an automatic float valve to maintain solution levels where required. Nutrient solution can be delivered by gravity or by a submersible pump, depending on the system layout. Periodically drain and clean the reservoir to remove any accumulated sludge or sediment resulting from impurities in the nutrient mix, as these can affect pump performance and nutrient balance.

**Environmental protection**

Climate is one factor you cannot control, but in a hydroponic garden you can modify or reposition your system to improve growing conditions. This may involve erecting a simple windbreak, adding a reflector to increase available light, warming the growing medium to promote faster germination, or using protective measures to guard against frost.

*Modular drip-irrigation system with expanded clay medium. Nutrient solution flows back to the reservoir by gravity.*

## Irrigation lines and cycle

For small beds, run a single 13 mm black polyethylene line along the length of the bed, with spaghetti lines spaced at about 300 mm intervals. In gravel systems, each plant should have its own feed line, while larger beds may require a main line along each row for even distribution. Emitters can be fitted to deliver approximately 4–6 litres per hour, or spaghetti lines can be inserted directly into the main feed line. In sand systems, a short length (about 50 mm) of 13 mm tube at the outlet helps distribute solution evenly around the root zone. Use black tubing and fittings throughout to minimise algae growth and maintain flow efficiency.

The system should be capable of delivering 6–10 litres per minute. When controlled by a timer, irrigation is typically applied two to five times daily, depending on plant maturity, weather, and seasonal conditions. In hotter weather, more frequent cycles may be required to maintain adequate moisture levels.

Test run-off solution weekly for total dissolved salts. If levels reach approximately 2000 ppm, flush the system with fresh water or a weak nutrient solution to prevent salt build-up and maintain healthy root conditions.

Uniform pressure across irrigation lines is essential to ensure each plant receives an equal supply of nutrient solution. Variations can result in uneven growth. Regularly inspect emitters and lines for blockages or wear, particularly where fine media or unfiltered water is used. Routine maintenance, such as flushing lines or replacing clogged emitters, will help maintain consistent performance.

Design the system so irrigation lines are easily accessible for cleaning, reducing downtime and simplifying maintenance.

*Coco peat drip-irrigation system. This high-efficiency method combines the excellent aeration and water-retention properties of coco coir.*

*Cinnamon basil propagated in various aggregate media in a flood-and-drain system. The tray is made from durable, high-impact ABS plastic.*

*Silverbeet grown in a trickle-feeding system over a shallow bed of fine-grade gravel.*

## Rockwool system

RRockwool systems may be operated as either free-drainage or recirculating systems. Owing to the high water-holding capacity of rockwool, irrigation can be less frequent than with many other media. However, good management practice is to apply nutrient solution several times daily, with frequency adjusted to suit the crop and growing conditions.

Frequent irrigation is required immediately after transplanting. As plants become established, five to ten irrigations per day are generally sufficient, increasing to as many as 20 per day during hot summer conditions. Irrigation should continue until approximately 10% of the applied nutrient solution drains from the slab, ensuring adequate flushing and preventing salt build-up.

**Rockwool system layout**

The layout for a recirculatory rockwool system is fundamentally the same as for a free-drainage rockwool system. It is necessary to collect what would otherwise be the drainage and to re-direct it back into the supply side of the irrigation system.

This can be achieved by setting slabs into a polyethylene line gully on the ground or by raising the slabs onto channels or benches from which the run-off solution is collected. In either case, the slope of the run should be regulated to give a reasonable water-to-air ratio within the slabs.

## Free drainage

For free-drainage systems, a level concrete floor with a slight fall towards the centre is ideal. This promotes effective drainage away from the slabs, improves light reflection, and supports good hygiene. An adequate catchment system should be provided to collect and remove excess nutrient solution from the growing area.

Soil or sand floors may also be used, provided they are carefully levelled, thoroughly disinfected, and covered with heavy-duty white plastic sheeting. This improves both light reflection and cleanliness.

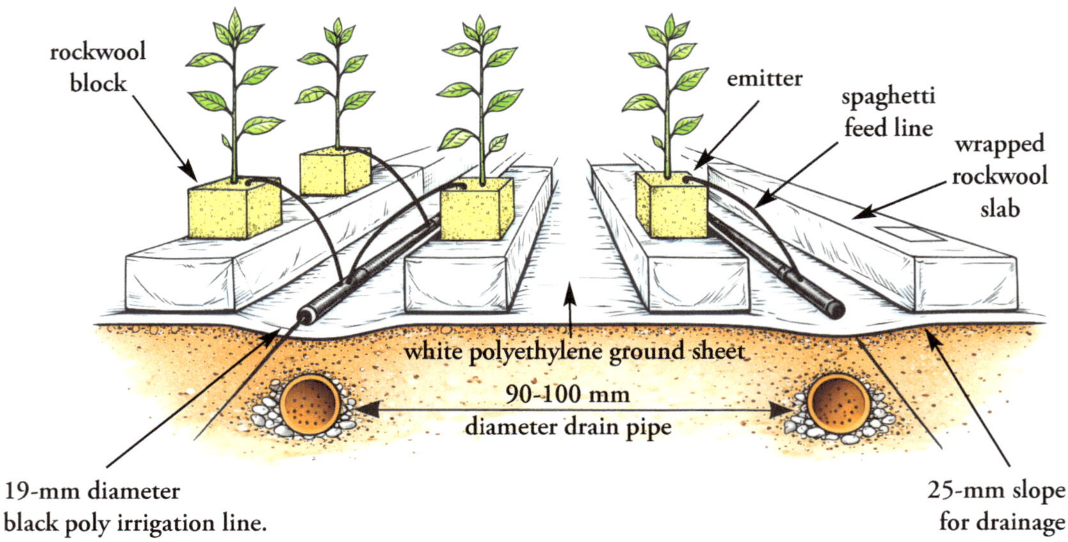


*A typical recirculatory rockwool system*

Rockwool beds should be spaced approximately 1 metre apart to allow easy access. Nutrient solution is delivered to the base of each plant via micro-irrigation lines (2–4 mm), supplied from a rigid PVC pipe or 19 mm poly line running along each bed and fed from a central nutrient reservoir.

*Tomato grown in a rockwool system using a black-and-white co-extruded polyethylene gully at ground level.*

## Recirculation

Recirculating rockwool systems are increasingly popular among home gardeners. While they use the same basic drip-irrigation principles as free-drainage systems, they differ in configuration and water management.

Systems may range from slabs placed on slightly inclined trays (typically around 1:200) to allow controlled drainage, to self-draining modular units. In all cases, the excess nutrient solution is collected, returned to the reservoir, and reused at regular intervals, improving efficiency and reducing water and nutrient loss.

## Aeroponic system

Aeroponic systems can be constructed using simple components such as a 20-litre bucket or deep tray, a submersible pump, an air pump with air stone, and a series of spray nozzles. The system operates by suspending plant roots within a closed chamber, where they are regularly misted with nutrient solution.

Spray nozzles should be positioned so that at least part of each plant's root system receives direct misting. The system may operate continuously at low pressure or intermittently, typically

*A-frame aeroponic system. Plants are supported in net pots, with roots suspended and periodically irrigated with a fine spray of nutrient solution.*

cycling on for 1 minute and off for 2 minutes. This intermittent operation improves aeration while maintaining adequate moisture levels around the roots.

An A-frame configuration is a practical variation, designed to maximise available space and light. Each side of the frame is fitted with a series of 50 mm holes to accommodate net pots holding the plants. For ease of operation, the frame should be hinged so that each side can be raised, allowing access for maintenance of the irrigation system, pump, and routine cleaning between crops.

## Electrical pumps

In hydroponics, pump selection is fundamental and depends on the type of system being used—whether flood-and-drain, NFT, aeroponic, or drip irrigation. Pumps are generally classified into two types: positive displacement and rotodynamic. Rotodynamic pumps are further divided into centrifugal, mixed-flow, and axial-flow designs.

In most hydroponic systems, rotodynamic pumps are preferred, with the centrifugal pump being the most widely used. These pumps are well suited to delivering large volumes of water at relatively low lift—ideal for small-scale systems. Rather than displacing a fixed volume of liquid, centrifugal pumps operate by hydrodynamic action. An impeller, enclosed within a volute chamber, spins to draw solution in through the inlet and discharge it through the outlet.

Centrifugal pumps are not self-priming and should therefore be installed below the level of the nutrient solution or fully submerged. This ensures reliable operation and prevents air locks within the system.

The main choice for most growers is between submersible and external (open-air) pumps. Submersible pumps are the most popular, as they are compact, quiet, efficient, and require minimal fittings. Their main limitation is constant exposure to nutrient solution, which can affect seals over time. For this reason, quality units are typically constructed from ABS plastic and fitted with chemically resistant seals and non-metallic components.

External pumps generally require more precise installation and additional fittings. Many must be primed before use or positioned so that the pump inlet remains flooded. Installing the pump below the reservoir level can overcome starting difficulties. Larger external pumps are often more robust and

**Electricity safety**

Electrical safety is critical in any powered hydroponic system, as water and electricity present a dangerous combination. Most systems rely on electrical equipment such as pumps and heaters, yet the growing environment itself—often humid, with damp surfaces—can increase the risk of electric shock from faulty appliances or supply.

For this reason, an automatic earth leakage circuit breaker (RCD) should be installed between the mains supply and all electrical equipment to provide essential protection.

easier to service, but are usually better suited to larger systems.

An alternative is the air-lift system, which uses an air pump to move solution by displacing water with air. While simple and inexpensive, it does not generate significant pressure and is therefore limited to systems with a single irrigation point. Its main advantage is that it simultaneously aerates the nutrient solution, increasing dissolved oxygen levels.

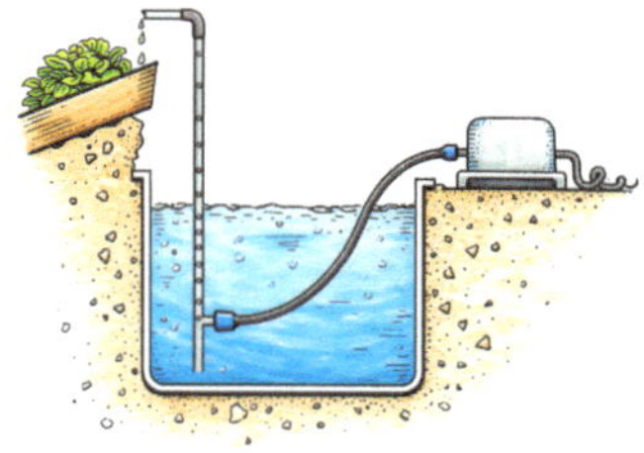

*Simple layout for an air-lift system.*

When designing an air-lift system, aim for efficiency by using consistent pipe diameters and minimising bends and fittings. This reduces resistance, improves flow, and helps maintain reliable operation.

## Filters

Filters are an essential component of any drip-irrigation system, preventing blockages and ensuring reliable operation. Inline plastic filters fitted with a fine mesh screen are commonly used and should be installed soon after the submersible pump. The screen should be removable for regular cleaning and maintenance.

As an alternative, or in addition to an inline filter, the pump can be placed inside a filter bag to trap debris before it enters the system. Commercial filter bags are available from hydroponic suppliers, though a simple nylon stocking can serve as an effective and inexpensive substitute.

*Layout for non-submersible pump, ensuring a flooded inlet.*

# Planting

Hydroponic systems may appear daunting at a commercial scale, but in practice they are relatively simple to construct and operate. Small-scale versions, based on the same principles, can be set up with minimal effort on a patio, rooftop, or in the backyard. For outdoor systems, planting should follow the same seasonal guidelines as soil-grown gardens. In controlled environments, however, crops can be produced year-round.

It is important to recognise that hydroponics does not overcome poor growing conditions. Factors such as temperature, light, and pest management remain critical to success. Hydroponically grown plants have the same basic requirements as those grown in soil; the principal difference lies in how the plants are supported and how nutrients are delivered.

## Seed germination

Large seeds can be sown directly into aggregate culture systems or simple do-it-yourself propagators. For smaller seeds, a more controlled medium such as rockwool or a perlite–vermiculite mix is recommended; a ratio of 3:1 has proven reliable for consistent germination. Once seedlings are well established and hardened, they can be transplanted into the hydroponic system. In water culture systems, however, plants should not be transferred until they have developed a sufficient root structure.

Seeds may be germinated in a range of media, including sand, perlite, vermiculite, rockwool, or coco peat. After sowing, maintain moisture by covering with a damp paper towel or sealing the container with plastic film to create a simple mini-greenhouse. Alternatively, a purpose-built propagator can be used. Once germination occurs, remove the cover and thin the seedlings as required. Use a dilute nutrient solution rather than plain water to moisten young plants, as most germination media provide little or no nutrition.

When seedlings are ready for transplanting, gently remove any adhering medium—except rockwool—from the roots by washing with tepid water. Take care to avoid damaging the fine root hairs, although small fragments of medium are not a concern. Seedlings raised in soil, including those purchased from nurseries, can also be transferred to hydroponics using the same method.

*Seeds can be planted in speedling cells (above) or rockwool blocks (below), which are then transplanted to the hydroponic system once the seeds have germinated.*

*Sealing the propagation container with plastic film creates a simple mini-greenhouse, maintaining humidity around the cuttings.*

Seedlings intended for most systems can be grown either bare-rooted or in rockwool or coco peat cubes or plugs. For water culture or aeroponic systems, it is preferable that roots protrude from the growing medium before transplanting. In channel systems, seedlings are typically established in rockwool cubes, coco peat plugs, or compressed perlite–vermiculite mixes. These substrates provide stability, support the plant, and exclude light from the root zone, helping to prevent root damage and inhibit algae growth.

*For woody plants, cuttings should be green-barked and not contain any of the woodier-type bark, which indicates age.*

*They should be 10-15 cm tall with three to five sets of well-developed leaves and a growing tip at the top.*

## Cloning

Cloning is one of the most efficient and reliable methods of plant propagation in hydroponics. It involves taking a cutting from a parent plant and growing it into a new plant—an example of asexual (vegetative) propagation. Unlike seed-grown plants, which result from sexual reproduction, cloned plants are genetically identical to the parent, provided they are grown under the same environmental conditions. Variations in light, temperature, or nutrition, however, can lead to differences in growth and form.

Cuttings can be taken at most stages of growth, but those taken from young, vegetative plants generally root more quickly and establish more vigorously. Cuttings from flowering plants tend to root more slowly.

To clone a plant, take a healthy cutting and dip the stem into a suitable rooting compound to reduce infection and encourage root development. Insert the cutting into an appropriate growing medium, such as those described earlier. For best results, select stems between 4 mm and 6 mm in diameter. Newly planted cuttings should be watered with a half-strength nutrient solution.

Healthy cuttings will retain most of their leaves, although some leaf loss during rooting is normal. Under good conditions, a strong root system should develop within about two weeks. After this period, increase feeding to full-strength nutrient solution. Careful water management is essential: overwatering can lead to root rot, while underwatering may cause the cuttings to dry out. Healthy roots are typically thick, white, and covered in fine root hairs; unhealthy roots tend to be thin, discoloured, and lacking in root hairs.

Once full-strength feeding begins, adjust the nutrient solution to a pH of approximately 6.0 to 6.5. When the cuttings have developed a sufficient root system, they can be transplanted into the hydroponic system and placed on a regular feeding regime. If grown in rockwool cubes or coco peat plugs, transplant them directly without removing the medium, whether into NFT, aggregate, rockwool, coco peat, or aeroponic systems.

**When to transplant**

Seedlings are ready for transplanting when they are about 1–2 cm high or have developed their first true leaves—those that resemble the mature plant, rather than the initial 'seed leaves' (cotyledons).

Seedlings raised under protected conditions are often too tender to be transferred directly into a hydroponic system. They should first be 'hardened off' by gradually exposing them to outdoor conditions. This process reduces transplant shock and helps maintain steady growth.

## Transplanting from soil

Plants can be transferred from soil to a hydroponic system with care. Gently wash away all soil from the roots using lukewarm water, taking care not to damage the fine root hairs. Success rates tend to be lower with mature plants, particularly those with well-developed taproots, as these are easily damaged during removal.

In hydroponic systems, plants develop a fine, fibrous root structure rather than large taproots. This reflects the ready availability of nutrients, removing the need for roots to extend and compete as they do in soil. As a result, younger plants adapt more readily to hydroponics. Plants with large, complex root systems often struggle to adjust and may suffer high mortality.

For this reason, it is generally preferable to propagate plants for hydroponics from cuttings rather than transplanting established soil-grown specimens.

## Nutrient solutions

Premixed nutrient solutions are readily available from local nurseries, hardware and specialist hydroponic stores throughout major and regional centres in Australia and New Zealand. They can also be purchased online from mail-order suppliers at comparable cost. These products are relatively inexpensive and simple to use. As you gain experience, you may choose to prepare your own nutrient solutions using raw components, which are also available from specialist outlets.

As plants absorb nutrients, they release by-products into the solution that can alter its chemistry, often causing the pH to rise. When the pH exceeds 7.0, it should be adjusted by adding small amounts of phosphoric or nitric acid to bring it back within the ideal range of 5.5 to 6.5. For larger plants, daily adjustment may be required. If the solution becomes too acidic, the pH can be raised using potassium hydroxide (KOH). A range of pH meters and colourimetric test kits are available to monitor solution levels accurately.

The nutrient solution should be replaced every two weeks when plants are small, and weekly once growth becomes more vigorous. Top up the system daily with water to maintain a constant volume. These are general guidelines that have proven effective, although practices may vary depending on the growing environment and the experience of the grower.

Responsible disposal of nutrient solution is important. Waste solution can be diluted and applied to soil-grown fruit trees, ornamentals, and garden beds, often with excellent results. In commercial closed systems, growers rarely discard the entire solution, instead bleeding off around 10 per cent of the total volume at regular intervals. In poorly maintained channel systems, minor leaks often result in a similar effect—albeit unintentionally.

*Artificial lighting allows vegetables, herbs and ornamentals to be grown indoors where natural sunlight is inadequate. Light intensity, duration and spectrum all influence plant growth and productivity.*

## Temperature

Plants grow best within a defined temperature range. Conditions that are too hot or too cold can lead to poor growth, abnormal development, and reduced yields. Warm-season vegetables and most flowering plants generally perform well between 15°C and 24°C. Cool-season crops, such as lettuce and spinach, prefer slightly lower temperatures, typically in the range of 10°C to 21°C.

## Water

Supplying water in hydroponic systems is generally straightforward in water culture, but can be more challenging in aggregate systems. During hot weather, a large tomato plant may use up to 2 litres of water per day. It is therefore essential to keep the growing medium adequately moist to prevent roots from drying out. If roots do dry, plants may recover once moisture is restored, but growth and yield are often reduced.

Water quality is equally important. Highly alkaline or salt-laden water, such as some bore supplies, can lead to nutrient imbalances and poor plant performance. Softened water may also contain excessive sodium, which is harmful to plants. As a general guide, water with a total dissolved solids (TDS) reading of 300 ppm or higher should be avoided, as it can disrupt nutrient uptake and overall plant health.

## Oxygen

Oxygen is essential for successful hydroponic growth. It supports root respiration and provides the energy required for water and nutrient uptake. In soil, oxygen is usually plentiful unless the soil is compacted. In hydroponic systems, however, plants growing in water can quickly deplete the available dissolved oxygen. To maintain adequate levels, aeration is commonly provided using an air stone placed in the nutrient reservoir.

**Oxygen**

Plants cannot survive without oxygen. This essential element supports respiration and is vital for healthy growth. In hydroponic systems, roots must have a continuous supply of oxygen to function properly.

If roots are deprived of oxygen—through constant immersion in poorly aerated water or compaction of the growing medium—they will suffocate and die.

## Light

All vegetable crops and many flowering plants require high levels of light. Hydroponically grown plants have the same requirements as those grown in soil, typically needing at least 8 to 10 hours of direct sunlight each day for good production. Artificial lighting is often a poor substitute, as many systems do not provide sufficient intensity for strong, economical growth.

High-intensity discharge (HID) lighting, such as metal halide and high-pressure sodium lamps, can deliver light levels in excess of 1,000 foot-candles and may be used successfully where natural sunlight is limited. However, the cost of fixtures and lamps can be significant, and may not be economically justified for the average home gardener.

## Plant spacing

Although hydroponic plants can be grown closer together than those in soil—since they do not compete for nutrients—plant density remains a critical factor in achieving optimum yields. For crops such as tomatoes and cucumbers, a density of two to three plants per square metre is generally recommended. While closer spacing is possible, it often results in reduced yields and poorer crop quality, primarily due to limited light penetration.

Wider spacing improves light distribution and promotes better air circulation, both of which contribute to healthier plants and higher-quality produce.

**Plant spacing**

Overcrowded plants quickly become tangled and shade one another, leading to weak growth and reduced yields. While experience is the best guide, some general spacing rules can be followed.

For larger crops such as tomatoes and cucumbers, allow no more than three plants per square metre. Smaller crops, such as lettuce, perform well at a spacing of 250–300 mm, providing adequate room for healthy development.

*Lateral pruning removes shoots at each leaf axil, directing growth along the support string and opening the canopy to improve light penetration.*

## Plant support

In soil, plant roots are surrounded and stabilised by the growing medium. In hydroponics, however, plants must be supported artificially. Common methods include trellises, A-frame structures, and wooden stakes, all available in various forms. For large fruiting crops such as pumpkins, rockmelons, and watermelons, individual fruits are sometimes supported using a mesh sling, such as a hair net or nylon stocking.

For ground-bed systems, a simple and economical support structure can be made using posts positioned at each end of the bed, with wires tensioned between them at different heights. This allows plants and vines to be supported progressively as they grow.

Another widely used method is the overhead support system, where a wire or cable is suspended above the growing bed. Plants are trained upward using string attached to the support, often secured with spools and clamps. This system is particularly suited to techniques such as layering, where older sections of the vine are defoliated after fruiting and the plant is gradually lowered to extend its productive life.

When using this method, the string should be tied to the overhead cable directly above each plant, allowing up to one metre of extra length for later adjustment. As the plant reaches the support height, it can be lowered as required. Clamps should be positioned beneath leaf petioles to provide effective support, and never above them. They should also be kept clear of flower clusters, as the weight of developing fruit may cause damage. For crops such as tomatoes and cucumbers, multiple clamps are required—typically spaced at intervals of around 30 cm—to ensure adequate support as the plant develops and fruits mature.

*Tall plants can be supported by string from an overhead wire or cable.*

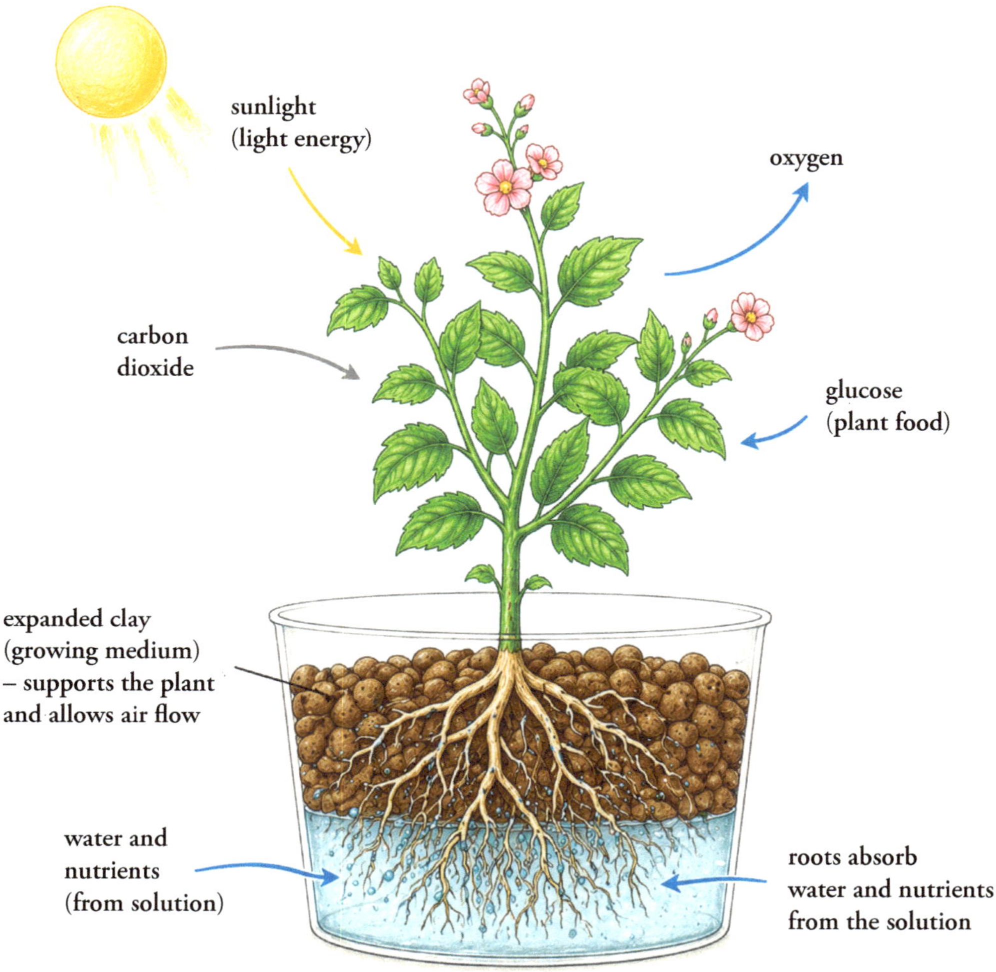


## Photosynthesis

Photosynthesis is the process by which plants use light energy to produce food. Using chlorophyll in the leaves, plants convert carbon dioxide from the air and water absorbed through the roots into glucose, releasing oxygen as a by-product.

This process is fundamental to plant growth. The glucose produced provides the energy required for development, while oxygen is released into the surrounding environment.

In hydroponic systems, the process remains the same as in soil-grown plants. The key difference is that water and nutrients are supplied in solution rather than drawn from the soil, ensuring they are readily available to support plant growth.

# Nutritional and environmental disorders

Nutritional disorders in hydroponics are caused by a deficiency or excess of one or more mineral elements. Affected plants exhibit symptoms specific to the element involved.

Mineral elements are classified as either mobile or immobile, although some may exhibit characteristics of both. Mobile elements—including nitrogen, phosphorus, potassium, magnesium and zinc—can be translocated from older leaves to actively growing regions of the plant. As a result, deficiency symptoms typically appear first on the lower, older leaves.

Immobile elements—such as sulphur, calcium, iron, boron, copper and manganese—cannot move readily within the plant. Deficiencies therefore develop in the younger leaves and growing tips, where demand is highest.

Early recognition of nutritional disorders is essential. The longer a deficiency or imbalance persists, the more widespread and less distinct the symptoms become, making diagnosis increasingly difficult. Common general symptoms include yellowing (chlorosis) and browning (necrosis) of plant tissue.

Toxic effects may arise from both essential and non-essential (trace) elements, including those present in water supplies such as sodium and chlorine. While plants can tolerate moderate excesses of major nutrients, trace elements must be carefully controlled, as even small imbalances can be harmful.

**Nutrient imbalance**
*Older cucumber leaves showing chlorosis and necrosis associated with nutrient imbalance, commonly seen with nitrogen deficiency.* (Image Dr Mary M. Peet)

An excess of one element can induce a deficiency of another, disrupting normal plant metabolism. For example, excessive nitrogen or phosphorus may reduce potassium uptake, while excess potassium can interfere with magnesium or calcium absorption. Similarly, a deficiency of one element may limit the plant's ability to accumulate others, leading to multiple imbalances. These interactions can make visual diagnosis difficult, particularly in the early stages.

A useful diagnostic technique is to grow an 'indicator' plant alongside the main crop. These plants are more sensitive to specific deficiencies and may show symptoms sooner. For example, when growing tomatoes, companion plants such as passionfruit, cucumber or lettuce can help indicate boron, calcium or magnesium deficiencies, respectively.

Early detection and correction are critical. Left unchecked, nutritional disorders reduce yield, impair fruit quality, and may ultimately lead to plant death.

In most cases, nutritional disorders appear simultaneously across a crop and progress from the lower leaves upward. In contrast, problems caused by pests or diseases typically begin on a few plants and spread gradually to others.

## Environmental problems

Symptoms of plant stress may also arise from environmental conditions. For this reason, suspected nutritional disorders should always be considered in the context of the overall growing environment, not solely in terms of the nutrient solution.

Temperature is a major factor. Plants will not grow optimally at low temperatures and may be damaged when conditions are too hot. Light is equally important; insufficient light results in weak growth and poor development. Atmospheric humidity also influences plant performance by affecting water loss from the leaves and overall moisture balance.

These environmental factors interact rather than act independently. Light, temperature and water availability all influence a plant's uptake and use of nutrients. For example, under low light conditions, plants generally require less nitrogen but relatively more potassium—a relationship of particular importance in glasshouse crops such as tomatoes.

**Guttation**

*Guttation in tomato. Excess root pressure causes droplets of xylem sap to exude from leaf margins, particularly overnight when humidity is high and transpiration is low.*

The interaction between light and nutrition can be observed in practice. A plant grown under normal light with insufficient nitrogen will show the typical pale green or yellowing associated with nitrogen deficiency. If the same plant is then shaded, the leaves may darken, masking the underlying nitrogen deficiency rather than correcting it.

Water uptake is also affected by environmental conditions. At low temperatures, root activity slows and water absorption is reduced, particularly if aeration is poor. This can lead to water stress, even when the growing medium appears adequately moist.

Plants respond in characteristic ways to unsuitable conditions. In low light, growth becomes elongated and weak, and leaves may lose their green colour. Excessively high temperatures can restrict growth, produce tougher, woody tissue, and give leaves a bluish appearance.

## Remedial action

Deficiencies or excesses of mineral elements can be recognised in a range of plant responses, including changes in leaf colour, density, size and shape; stem thickness and internode length; root colour and structure; and the abundance, timing and quality of flowers and fruit. Careful observation of these signs is the key to diagnosing nutritional disorders.

When a disorder is recognised—or even suspected—the first step is to replace the nutrient solution with a fresh, correctly balanced mix. Where a deficiency has been identified, a suitable foliar spray may be applied to provide a rapid response. Care should be taken to avoid excessive concentrations, which can damage plant tissue. It is advisable to treat a small number of plants first and observe the results over several days before applying the treatment to the entire crop. A visible response may take 7 to 10 days.

In cases of nutrient excess or toxicity, the growing medium should be flushed with fresh nutrient solution to reduce residual salt levels. This may need to be repeated over several days, depending on the severity of the problem. Plain water should not be used, as it can disrupt plant membranes and deprive roots of essential nutrients.

Regular monitoring of pH, electrical conductivity (EC), nutrient solution temperature and water quality can help prevent many nutritional problems before they become serious. Maintaining accurate records of nutrient additions, crop performance and any corrective actions taken provides a useful reference for future crops and can assist in identifying recurring issues associated with seasonal conditions, water supply or management practices.

**Tipburn**

*Tipburn is commonly associated with calcium deficiency, often exacerbated by high temperatures. It is easily recognised by the inward curling of young inner leaves, which develop a scorched or necrotic appearance along the margins. In root vegetables, calcium deficiency may also present as distorted growth, with roots appearing forked or bent.*

# Deficiency symptoms

Symptoms usually appear first in older or younger leaves depending on mobility. While symptoms vary from plant to plant, the general symptoms of nutrient deficiencies can be listed as follows:

*Nitrogen deficiency*
(Image Dr William R. Jarvis)

**Nitrogen**

Chlorosis (yellowing) of the whole plant, often with reddening.
Older leaves are usually affected first.

**Potassium**

Older leaves may show necrotic spots or marginal burn.
Younger leaves may develop red pigmentation.
Interveinal chlorosis remains in the inner leaf, with veins still somewhat green.
Leaves appear dry, brittle, and curled, especially along the margins.

*Phosphorus deficiency*
(Image NSW DPI)

**Phosphorus**

Leaves or petioles may develop red or purple pigmentation.
The colour change is broad and uniform, not patchy like disease or pest damage.
Leaves appear slightly thickened and dull.
Typically shows first on older leaves, as phosphorus is mobile in the plant.

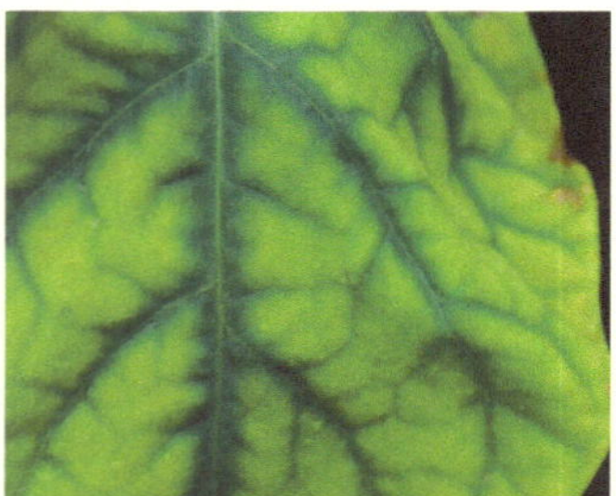

*Magnesium deficiency*
(Image licensed from Shutterstock)

**Sulphur**

Chlorosis of the whole plant.Younger leaves are often affected first.

**Magnesium**

Marginal or interveinal chlorosis, often strongly coloured.
Green areas may form an 'arrowhead' pattern in woody plants.
Strong reddening may border the chlorotic zone, on older tissue first.
Leaf tips and margins often turn or cup upwards.

*Iron deficiency*

**Calcium**

Death of the growing point.
In fruit crops, disorders are species-specific (e.g.blossom-end rot in tomatoes and capsicums).
In leafy crops, such as lettuce, tipburn is common.

**Iron**

Interveinal chlorosis.
In severe cases, young leaves may bleach, followed by necrosis.
Symptoms occur on young leaves first.

**Chlorine**

Chlorosis followed by necrosis, often accompanied by wilting.

Leaf tip and margin burn may occur.

Leaf bronzing and mottling.

Roots may appear stunted and thickened near the tips.

**Manganese**

Interveinal chlorosis.

In severe cases, necrotic spots or streaks may develop.

Symptoms often begin in middle leaves.

Moderate deficiency — mottled yellow/green pattern

Severe interveinal chlorosis — bright yellow between veins, veins remain green

*Manganese deficiency*
(Image Haifa Group)

**Boron**

Death of growing points.

Axillary buds may develop, producing a witch's-broom appearance.

Leaf distortion may occur in some species (e.g. grape).

Cracking and corky tissue around the stem (calyx).

Tissue appears dry, rough, and necrotic, rather than soft or rotting.

Petiole cracking (celery).

Internal hollowness in some root vegetables may also occur.

*Boron deficiency*
(Image NSW DPI)

**Zinc**

Small leaves and rosetting.

Early release of older leaves.

Leaf tips will look burned.

In milder cases, a chlorotic mottle may be evident.

Interveinal chlorosis with veins staying green, but the pattern is broader and more banded than manganese.

**Copper**

Death of young leaves.

Chlorosis may occur.

Failure of fertilisation and fruit set is common.

S-shaped shoot growth and fruit gumming may occur in citrus.

*Copper deficiency*
(Image Haifa Group)

**Molybdenum**

General pallor in legumes.

In non-legumes, a mottled pale appearance is evident.

In mature leaves, marginal burn may occur (e.g. rockmelon, maize, sunflower).

Whiptail may develop in cauliflower.

*Molybdenum deficiency*
(Image NSW DPI)

# Toxicity symptoms

Deficiencies typically follow structured patterns related to leaf age or vein structure, whereas toxicities more often present as burning, spotting, or general tissue collapse affecting multiple leaves simultaneously. Symptoms are also given for fluorine and aluminium since these elements are often found in water supplies.

*Nitrogen toxicity*

**Nitrogen**

Leaves turn a deep, almost shiny green.
The (nitrate) root system is restricted.
In the case of ammonium nitrogen, there is initial chlorosis,..

*Phosphorus toxicity*
(Image Haifa Group)

**Phosphorus**

Interveinal chlorosis in younger leaves resemble iron deficiency.
Necrosis and tip dieback may follow in susceptible plant species.
Marginal scorching and shedding of older leaves.

**Potassium**

Symptoms may seem similar to those of magnesium deficiency.
Possible manganese, zinc and iron deficiency.

*Sulphur toxicity*
(Image Haifa Group)

**Sulphur**

A reduction in growth and leaf size.
Occasional interveinal chlorosis or leaf burning.

**Magnesium**

Stunted roots.
Copper-colored leaves and premature senescence.

**Calcium**

In practice, calcium toxicity often looks like a combination deficiency problem, not a single clear symptom.
Excess calcium interferes with uptake of other nutrients

*Chloride toxicity*
(Image NSW DPI)

**Iron**

Bronzing or dark brown spots on lower leaves.
Stunted roots, reduced growth.

**Chloride**

Bronzing, chlorosis and marginal burn.
Leaf drop may be premature.

**Manganese**

Chlorosis, beginning at the leaf edge of older leaves, sometimes with an upward cupping. Interveinal bronze yellow chlorosis in beans.
Orange-yellow marginal and interveinal chlorosis in lemons
Brown tar spots in orange leaves
Necrosis in apple bark.

*Manganese toxicity*
(Image Haifa Group)

**Boron**

Interveinal necrosis.
Stunted growth.
Leaf chlorosis (yellowing).
Dark necrotic spots, starting on older leaves.

**Zinc**

Chlorosis (Yellowing): Pale yellow or white interveinal chlorosis, particularly on young leaves.
Necrosis and Spotting: Dark pigmented spots, blotches on older leaves, or red/purple pigmentation on petioles and leaf margins.
Severely inhibited, thick, or brownish roots.
General wilting of the plant.

*Zinc toxicity*
(Image NSW DPI)

**Copper**

Reduced growth followed by symptoms for iron chlorosis.
Stunting and reduced branching are sometimes apparent.

**Molybdenum**

Tomato leaves turn golden yellow.
Cauliflower seedlings turn bright purple.

*Boron toxicity*
(Image Haifa Group)

**Fluorine**

Scorching of leaf tip and margin, extending into interveinal areas.
Chlorosis: Yellowing of leaves.
Reduced yield and distorted fruit.

**Aluminium**

Symptoms in shoots are frequently similar to those for phosphorus deficiency, indicating an impairment of phosphorus absorption and metabolism by high levels of aluminium.
Roots grown in high levels of aluminium are frequently short, with many short laterals.
Root tips are commonly brown.

*Salt toxicity*
(Image NSW DPI

# Commonly confused disorders

## Potassium deficiency vs Salt (EC) toxicity

High salt levels restrict water uptake and can mimic deficiency symptoms. Always check EC before adjusting nutrient ratios.

*Potassium deficiency*
(Image licensed from Shutterstock)

**Potassium deficiency**

Yellowing begins at leaf margins
Develops into brown edge scorch
Damage moves inward
Older leaves affected first
Veins remain visible; pattern is structured
Progression is gradual

*Salt toxicity*
(Image NSW DPI)

**Salt toxicity (high EC)**

Tipburn and general leaf scorch
Damage appears more uniform across the leaf
Early stages may show a pale or burnt edge ("halo")
Tissue appears dull, dry, collapsed, or brittle
Progression is rapid and widespread

***Diagnostic rule***

Edge burn moving inward = potassium deficiency
Rapid, widespread scorch = salt toxicity
Always check EC before increasing potassium

## Calcium deficiency vs Tipburn / Blossom End Rot

Tipburn and blossom end rot are usually caused by poor calcium transport, not a lack of calcium in the nutrient solution. Check watering, EC, and airflow before adding calcium.

*Tipburn*
(Image licensed from Shutterstock)

**Calcium deficiency (true nutrient deficiency)**

New leaves: distorted, twisted, malformed.
Roots: stubby, poor development
Fruit: general deformity (not just end rot)

**Tipburn / Blossom End Rot (physiological)**

Occurs on otherwise healthy plants

**Lettuce (tipburn)**

Inner young leaves affected
Margins scorched and necrotic
Leaves curl inward

*Blossom end rot*
(Image Eyal Ronen)

**Tomato / Capsicum (BER)**

Blossom end of fruit
Dark, sunken lesion
Tissue becomes leathery, then rots

***Diagnostic rule***

Calcium deficiency = insufficient calcium available to the plant
Tipburn / BER = failure of calcium transport
Distorted new growth = calcium deficiency
Burnt tips or fruit ends on an otherwise healthy plant = transport failure

## Magnesium deficiency vs Potassium deficiency

High potassium levels can induce magnesium deficiency. Always check overall nutrient balance before adjusting.

*Magnesium deficiency*
(Image Mary M. Peet)

### Magnesium deficiency

Interveinal chlorosis (yellowing between veins).
Veins remain green and clearly defined
Pattern is structured and vein-related.
May develop reddish or purple margins in some crops.
Advanced stages show necrotic spots, not edge burn.

### Potassium deficiency

Yellowing begins at leaf margins.
Develops into brown edge scorch.
Damage moves inward from the edges.
Leaf appears dry and brittle
Pattern is edge-based, not vein-based.

*Potassium deficiency*
(Image licensed from Shutterstock)

*Diagnostic rule*

Yellow between veins = magnesium deficiency
Burnt edges moving inward = potassium deficiency

## Iron deficiency vs Manganese deficiency

Both iron and manganese availability are strongly influenced by pH. Iron becomes unavailable at high pH; manganese availability also declines as pH rises.

*Iron deficiency*
(Image Haifa Group)

### Iron deficiency

Bright yellow to almost white leaves
Veins remain sharp and dark green
High contrast: "yellow leaf with green skeleton"
Appears on youngest leaves first
Growth slows; severe cases show bleaching

### Manganese deficiency

Mottled yellowing (less intense)
Veins remain green but less sharply defined
Pattern is fine, net-like or speckled
May develop small necrotic specks later
Affects young to mid leaves; not strictly age-dependent

*Manganese deficiency*
(Image NSW DPI)

*Diagnostic rule*

Bright yellow with sharp green veins = iron deficiency
Mottled yellow with softer veins = manganese deficiency

*Nitrogen deficiency*
(Image Dr William R. Jarvis)

## Nitrogen deficiency vs Low light (environmental)

Before increasing nitrogen levels, check light conditions. Poor light reduces growth and can cause pale plants even when nutrients are adequate.

### Nitrogen deficiency

Older leaves yellow first
Chlorosis progresses upward
Growth is slow, thin, and weak
Reduced flowering and fruiting
Reduced new leaf growth

*Low light*
(© Petrus78 | Stock Free Images)

### Low light

Whole plant looks uniformly pale
Leaves often larger but thinner
Stems elongated (stretching)
No clear leaf-age pattern

### *Diagnostic rule*

Older leaves yellow first = nitrogen deficiency
Whole plant pale and stretched = low light

*Botrytis*
(Image Dr Mary M. Peet)

## Disease vs Nutritional disorder

If symptoms appear as distinct spots or lesions, do not adjust the nutrient solution. Check for disease first.

### Disease

Defined lesions (spots with edges)
Often circular or repeating shapes
May have halos or centres
Damage is patchy and localised

*Iron deficiency*

### Nutritional disorder

Patterns, not spots
Symptoms follow a consistent pattern across the leaf or plant
Follows:
- veins
- margins
- leaf age

Symptoms follow a consistent pattern across the plant
No clean lesion boundaries

### *Diagnostic rule*

Spots with defined edges = disease
Patterns without defined spots = nutritional

# Managing pests and diseases

## Disease control

Infectious diseases can be the nemesis of any gardener. Many cannot be eradicated and, at best, can only be contained, so prevention is critical. While disease problems are less common in hydroponics than in soil, they still pose a serious threat if allowed to develop unchecked. With increasing awareness of the risks associated with chemical treatments, home gardeners should maintain strict sanitation to prevent and limit the spread of disease.

Even with good preventative practices, disease organisms can still gain a foothold in a hydroponic crop. Pathogens may be introduced on contaminated seedlings or cuttings, or carried in water from trays, pots or growing systems. Some diseases require moisture on the foliage to develop, while others thrive in excessively wet root conditions. Certain diseases, such as rusts and botrytis, produce airborne spores that can spread from nearby plants.

Hydroponic gardeners should inspect their plants daily so that problems are detected early and control measures can be applied promptly.

Plant diseases are caused by four main groups of organisms — viruses, bacteria, fungi and nematodes. Understanding how these pathogens spread and develop is essential for effective control.

Early detection is the key to control — once disease is established, options are limited.

**Pests and diseases**

In hydroponics, soilborne pests and diseases are largely eliminated. However, plants grown in soilless systems remain vulnerable to a range of common problems. Vigilance and early detection are essential for effective control. The key to a healthy system is cleanliness, good environmental control and sound growing practices.

## Viruses

Viruses are microscopic organisms that live within plant cells and cause a wide range of abnormalities. Unlike animals, plants do not produce antibodies, so they do not recover from viral infections or develop immunity. Some resistance can be bred into plant varieties, but once infected, a plant remains infected for life, even if symptoms temporarily fade or appear to disappear.

There are no chemical treatments available to control viral

*Lettuce ring necrosis disease, caused by a virus transmitted through nutrient solutions by the fungus Olpidium virulentus.*
(Image Dr Len Tesoriero)

diseases. Infected plants act as a source of infection and can spread the virus throughout the system. The only effective control is to remove and destroy affected plants promptly.

Viral diseases commonly cause deformity, stunting and dwarfing. Leaves are usually the first to show symptoms, often displaying mottling, streaking, blotching or ring patterns in shades of light green, yellow, brown or black. Leaves may also become distorted, puckered or curled. Flowers can be reduced in size, misshapen, discoloured or, in some cases, develop into leafy structures.

Viruses are not usually transmitted through seed, but early infection of young plants can lead to significant crop loss. While viruses can spread within a plant from cell to cell, they require a carrier to move between plants. This is most commonly through insect feeding, particularly by sap-sucking pests such as aphids. Viruses may also be spread by grafting and contaminated tools, making good hygiene essential.

Remove infected plants immediately — viruses cannot be cured and will spread if left in the system.

## Bacteria

Bacteria are single-celled micro-organisms that can be difficult to control once established. Although a limited number of bactericides are available, effective control relies mainly on prevention and the prompt removal of infected plants.

Bacterial diseases often develop under warm, moist conditions and spread through water, handling, or contaminated tools and equipment. Good hygiene and careful watering practices are therefore essential to limit their spread.

Common bacterial diseases include bacterial wilt (*Pseudomonas caryophylli*), which affects carnations ; bacterial spot (*Xanthomonas spp.*) and bacterial speck (*Pseudomonas syringae pv. tomato*), which commonly affect tomatoes and capsicums; bacterial leaf spot of lettuce (*Xanthomonas campestris pv. vitians*); angular leaf spot of cucurbits (*Pseudomonas syringae pv. lachrymans*); and soft rot (*Erwinia chrysanthemi*), which commonly affects cuttings, bulbs and soft plant tissue. Other important diseases include bacterial blight (*Xanthomonas pelargonii*) in pelargoniums and crown gall (*Agrobacterium tumefaciens*), which infects a wide range of ornamental plants including roses and chrysanthemums.

Avoid wet foliage and handle plants carefully — bacteria spread easily through water, contact and contaminated tools.

**Bacteria prevention**

The most effective control is prevention. Remove infected plants promptly and sterilise pots, containers, channels and irrigation equipment. Between crops, clean systems thoroughly using a mild chlorine solution (household bleach), then flush well with fresh water before replanting.

## Fungi

Fungal diseases are the most common of the four groups and have been the downfall of many a home gardener. However, they are generally the easiest to recognise and respond well to control measures.

Fungi are more complex than bacteria, being multi-cellular organisms that often consist of several tissues. Most reproduce asexually by forming spores. The most easily recognised fungal disease is powdery mildew, characterised by a whitish powder on the surface of leaves, stems and sometimes petals. Spores are easily detached and carried on air currents to surrounding plants, where they initiate new infections.High humidity favours mildew development, so ventilation and temperature control in greenhouses should be adjusted to avoid damp, stagnant conditions.

To control mildew, grow resistant varieties and use appropriate fungicides. Sulphur-based products remain widely used. Dusting sulphur can be applied quickly and easily, and preventative treatments may be applied at regular intervals. However, care should be taken when using sulphur in hot weather or enclosed conditions, as it may cause leaf damage.

Botrytis (*Botrytis cinerea*) is a common grey mould affecting a wide range of ornamental and vegetable plants. It is identified by fuzzy, grey spore masses on infected tissue, although these may not be visible under dry conditions. Spores are easily spread by air movement but require moisture to germinate, typically infecting petals, damaged tissue or senescing plant material.

High humidity also favours botrytis and should be controlled through good ventilation. Avoid wetting foliage and minimise condensation. Because the fungus readily colonises dead or dying tissue, strict sanitation is essential. While fungicides can assist, effective control depends largely on environmental management.

Root diseases such as rhizoctonia, pythium, thielaviopsis and colletotrichum cause damping-off in seedlings and root and basal stem rots in established plants. These organisms are widespread and can infect a broad range of crops.

Pythium, though not a true fungus, behaves similarly and is commonly grouped with fungal diseases in hydroponics. It is a waterborne organism that spreads readily through nutrient solutions and thrives in poorly aerated, wet conditions.

Symptoms of pythium include a soft, water-soaked rot of the roots, which become brown, hollow and collapse easily. Affected plants may wilt rapidly despite adequate moisture.

Poor aeration and warm nutrient solutions are the most

*Tomato fruit affected by fusarium rot, showing sunken lesions and white fungal growth typical of this destructive disease under warm, humid conditions.*
(Image Dr William R. Jarvis)

### Fungi prevention

The most effective control is prevention. Maintain good sanitation, remove diseased plant material promptly, and use resistant varieties where possible. Good airflow and humidity control are essential. Where treatment is required, sulphur and copper products are commonly used, with sulphur generally preferred for low-toxicity applications.

*Botrytis cinerea is a necrotrophic fungus that feeds on dead or damaged tissue and affects many plant species, including strawberry. Commonly known as grey mould, it produces a characteristic grey, fuzzy growth on infected fruit and plant surfaces.*
(Image licensed from Shutterstock)

*Mirafiori lettuce big-vein virus (now known as Ophiovirus mirafioriense). It is transmitted to roots by the fungus Olpidium virulentus.*
(Image Dr Len Tesoriero)

common causes of pythium outbreaks in hydroponic systems.

Rhizoctonia produces a drier, brown rot and is favoured by moderate moisture conditions. It often develops at higher temperatures. Thielaviopsis causes lesions that quickly turn black due to the production of large numbers of dark spores. This disease is less likely to occur under acidic conditions (pH 4.5–5.0). Colletotrichum is frequently associated with root decay in NFT systems and can invade otherwise healthy tissue.

These root pathogens spread through contaminated media, plant material, tools and nutrient solutions.

Preventative measures are similar for all root diseases. The most effective controls include the use of well-drained media, disinfection or pasteurisation of growing media, sterilisation of trays, pots, tools and benches, the use of clean planting material, and strict hygiene throughout the system.

Verticillium diseases affect a wide range of ornamental plants, including chrysanthemums, snapdragons, roses, geraniums and begonias. Symptoms vary with the host. In chrysanthemums, marginal wilting is followed by chlorosis and eventual leaf death. Snapdragons may appear healthy until flowering, when sudden wilting occurs. In roses, buds may fail to open and foliage becomes mottled before decline. Begonias often show yellowing at the leaf margins, with a distinctive glossy underside.

As with other fungal diseases, good hygiene and sourcing healthy plants are essential. However, Verticillium is rarely a problem in well-managed hydroponic systems.

## Nematodes

*Large galls of the root knot nematode* Meloidogyne incognita *on tomato.*
(Image William R. Jarvis)

Nematodes, also known as eelworms, are microscopic parasites that live in soil, hydroponic substrates and water, and feed on plant tissue. They are elongated, thread-like organisms, usually tapered at both ends, and are rarely visible to the unaided eye.

While many nematodes are harmless and form part of the natural ecosystem, some species can damage plants, particularly when populations build up. In hydroponics, nematode problems are less common than in soil, but they can still occur in contaminated media or planting material.

Harmful nematodes can be grouped into two main types: those that feed on roots, and those that attack leaves and buds.

Root-knot nematodes are among the most damaging. Infected plants are often stunted and may wilt during warm conditions. The most obvious symptom is the formation of galls or swellings on the roots. Although affected plants may continue to grow and produce under good nutritional conditions, heavy infestations will reduce vigour and yield.

Leaf and bud nematodes cause distortion of leaves, buds and fruit, particularly in ornamentals and strawberries. Leaf infections typically begin as small yellow or brown spots on lower leaves, which may enlarge, darken and lead to defoliation under favourable conditions.

Control of nematodes relies on prevention. Infected plants should be removed and destroyed to prevent spread. Growing media should be pasteurised or replaced, and all equipment, trays and tools thoroughly cleaned and sterilised before reuse.

Nematodes are best controlled by hygiene—once established, they are difficult to eliminate.

**Nematode prevention**
The most effective control is prevention. Remove and discard infected plants, and sterilise or replace the growing medium before replanting. When in doubt, replace the medium to avoid reinfestation.

## Disease prevention

Disease prevention relies on a consistent programme of good sanitation and environmental control. Many plant pathogens, such as botrytis, require free moisture on plant surfaces for spores to germinate.

Free water often forms as condensation. Warm air holds more moisture than cool air, so as temperatures drop—particularly at night—moisture condenses on plant surfaces once the dew point is reached.

In greenhouse environments, condensation can be reduced by:maintaining adequate ventilation, circulating air with fans periodically exchanging air in enclosed spaces.

High humidity also encourages diseases such as powdery mildew. This can be controlled by improving airflow and ventilating regularly. Watering early in the day allows excess moisture to evaporate as temperatures rise, reducing the period of leaf wetness.

Root rot and damping-off diseases are commonly spread through contaminated media, water and plant material. Minimising splashing during watering, particularly by using automated systems, helps reduce the spread of pathogens. Good drainage is essential, as many root diseases are favoured by saturated growing media. Waterlogging of the root zone should be avoided at all times.

Strict hygiene is critical. Trays, pots and containers should be cleaned and sterilised before use. Tools, supports and irrigation systems must also be kept clean. Remove plant debris promptly and dispose of it away from the growing area to prevent reinfection. Never leave prunings on the floor, where pathogens

**Disease prevention**
- Purchase disease-free plants from reputable nurseries
- Clean and sterilise equipment regularly
- Inspect plants frequently for early symptoms
- Remove plant debris promptly from the growing area
- Water carefully to minimise free moisture on plant surfaces

**Insect control**
Maintain a balance between pests and beneficial insects. Predators and parasites help keep pest populations at manageable levels, reducing the need for sprays or other controls.

can easily spread.

When purchasing plants, select healthy stock from reputable nurseries and inspect carefully for any signs of disease. If propagating your own plants, use clean equipment, maintain good hygiene and monitor plants regularly so that problems are detected early.

Most diseases begin with excess moisture and poor hygiene — control these, and problems are greatly reduced.

## Pest control

Insects, mites and other animal pests pose a continual threat to the quality of hydroponic crops. An effective pest management programme should therefore focus on prevention, early detection and timely control.

*Early spider mite damage appears as fine yellow stippling caused by sap removal from leaf cells. Heavy infestations lead to bronzing, necrosis and premature leaf senescence.* (Image licensed from Shutterstock)

While a range of chemical sprays is available, many hydroponic gardeners prefer biological and organic methods. The use of broad-spectrum pesticides can disrupt the natural balance of beneficial and harmful insects, often leading to further problems.

A healthy garden supports a balance between pest species and their natural predators. Control measures should aim to reduce pests to manageable levels rather than eliminate them.

Pests are commonly introduced on nursery plants, which should be inspected carefully before entering the growing area. They may also be carried in by wind or attracted from neighbouring gardens. Early detection is critical—small infestations are far easier to manage than established ones.

Pests often establish in sheltered or favourable microclimates, such as warm, protected areas or dense plant growth. Regular inspection of these areas is essential, with particular attention to the undersides of leaves where many pests feed and breed.

Some pests, such as slugs and snails, hide in damp, shaded locations or beneath containers and growing media, emerging at night to feed. Understanding where pests hide and recognising early signs of damage are key to effective control.

Inspect plants regularly—early detection is the most effective form of pest control.

**Preventing insect problems**
Healthy plants are more resistant to pest attack. Maintain balanced nutrition to support strong, resilient growth. While nitrogen is essential, excessive levels can produce soft, succulent growth that encourages sap-sucking insects. Use a well-balanced nutrient solution and avoid overfeeding.

## Spider mites

Also known as two-spotted mites or red spider mites, spider mites are a common and persistent pest in hydroponic systems. They are extremely small and often go unnoticed until damage appears.

Early symptoms include fine yellow speckling on the upper surface of leaves. As feeding continues, leaves may become bronzed, dry and weakened. On close inspection, tiny oval mites can be seen on the undersides of leaves. A fine webbing over plant surfaces is a characteristic sign and can be revealed by misting plants with water, which causes droplets to cling to the threads.

Spider mites are easily spread, often carried on clothing or introduced by air movement. They are most active in hot, dry conditions, particularly in enclosed or poorly ventilated environments. At lower temperatures they become inactive, but populations increase rapidly as temperatures rise.

Because spider mites prefer dry conditions, increasing humidity and regularly misting plants can help suppress infestations. Maintaining good airflow and monitoring plants closely are essential for early detection and control.

Fine speckling and webbing on leaves are early signs of spider mite—act quickly before populations build.

*Adult whitefly* (Trialeurodes vaporariorum), *a common sap-sucking pest.* (Image Marilyn Steiner, reproduced with permission NSW DPI)

## Whitefly

Whitefly is a common pest, easily recognised by its small, white, moth-like appearance. Adults are typically found on the undersides of leaves, where they feed by sucking sap from the plant.

When disturbed, whiteflies rise in a small cloud before settling back onto the foliage. Heavy infestations cause yellowing, reduced vigour and poor plant growth.

Whitefly populations develop rapidly in warm, sheltered conditions. Eggs and immature stages are usually found on the undersides of leaves, making regular inspection essential.

Biological control using the parasitic wasp *Encarsia formosa* is highly effective. The parasite lays its eggs inside whitefly nymphs, preventing them from developing into adults.

Yellow sticky traps can also be used to monitor and reduce populations, as whiteflies are strongly attracted to the colour yellow.

A sudden cloud of small white insects when disturbed is a clear sign of whitefly infestation.

*Lettuce aphid* (Nasonovia ribis-nigri), *an invasive pest that colonises deep within lettuce hearts, causing distortion and contamination of marketable crops.* (Image NSW DPI)

### Aphids

Aphids are a common and potentially damaging pest. Early signs include leaves that are curled, puckered or discoloured. On closer inspection, dense colonies of small, soft-bodied, pear-shaped insects can be found on tender growing tips and the undersides of leaves.

Aphids feed by sucking sap from plant tissue, weakening

*Colony of aphids feeding on plant sap. Heavy infestations can stunt growth, distort foliage and spread viral diseases.* (Image Marilyn Steiner, reproduced with permission NSW DPI)

growth and reducing vigour. They also transmit plant viruses, making them one of the more serious pests in hydroponic systems. In addition, aphids produce sticky honeydew, which can lead to the development of sooty mould on leaf surfaces.

Aphid populations can increase rapidly, particularly in warm, protected environments. Individuals vary in colour, ranging from green and yellow to black, depending on the species.

Natural predators such as ladybirds and green lacewings are highly effective in controlling aphid populations. Early detection and prompt action are essential to prevent rapid spread.

Curled leaves and sticky residue are early signs of aphids — check growing tips and leaf undersides.

## Fungus gnats

Fungus gnats are small, dark, mosquito-like flies commonly found in hydroponic systems, particularly those using organic or moist growing media. While the adults are mostly a nuisance, the larvae can cause significant damage by feeding on roots and young plant tissue.

Adult fungus gnats are often seen flying close to the growing medium or resting on plant surfaces. The larvae live in the upper layers of moist media, where they feed on organic matter, algae and fine roots. Heavy infestations can weaken plants, reduce growth and increase susceptibility to root diseases such as pythium.

Fungus gnats thrive in damp conditions and are most common where media remains constantly wet or where algae is present. Overwatering and poor drainage greatly increase the risk of infestation.

Control relies on reducing moisture levels in the growing medium and improving drainage. Allowing the surface of the medium to dry between watering can significantly reduce larval survival. Yellow sticky traps are useful for monitoring adult populations. Good hygiene, removal of algae and avoidance of overwatering are essential for long-term control.

Fungus gnats indicate overly wet conditions — correct the environment first.

*Adult fungus gnat* (Sciaridae). *While adults cause little direct damage, larvae feed on roots and organic matter and can transmit root diseases.*
(Image licensed from Shutterstock)

## Thrips

Thrips are small, slender insects that feed by rasping the surface of leaves, flowers and young fruit, then sucking up the released cell contents. Their feeding causes a distinctive silvery or bronzed appearance on leaves, often accompanied by fine black specks of excrement. New growth may become distorted, and flowers can be scarred or fail to develop properly.

These pests are particularly troublesome in warm, protected environments such as greenhouses and hydroponic systems, where populations can build rapidly. Thrips are highly mobile and easily spread between plants, making early detection important. They are often found on the undersides of leaves or within flowers, where they are protected from casual inspection.

In addition to direct feeding damage, thrips are known vectors of plant viruses, which can significantly reduce crop quality and yield. For this reason, even low populations should not be ignored.

Monitoring is best carried out using blue or yellow sticky traps, combined with regular inspection of young leaves and flowers. Biological control using predatory insects can be effective, particularly when introduced early. Good hygiene, removal of heavily infested plant material, and exclusion of pests from the growing area all help to reduce the risk of infestation.

As with most pests, consistent observation and prompt action are the keys to successful control.

*Western flower thrips* (Frankliniella occidentalis) *feeding on plant tissue. Thrips cause silvering, scarring and distortion of leaves and flowers and can transmit plant viruses.* (Image Marilyn Steiner, reproduced with permission NSW DPI)

## Ways to control insect pests

Non-chemical pest control methods fall into two main categories: biological and organic.

Biological control involves the use of living organisms to reduce pest populations. These include natural predators, parasites and microbial agents that attack insects. Simple methods such as hand removal and the use of companion plants can also help deter pests. In controlled environments, growers may introduce 'banker' plants to support beneficial insects and improve long-term pest management.

Banker plants are non-crop plants introduced into a growing area to support beneficial insects. They provide a habitat or alternative food source for predators or parasites, allowing these beneficial organisms to establish and remain active even when pest levels are low.

Organic control refers to the use of low-toxicity, non-synthetic treatments and cultural practices. These include plant-based sprays, careful crop management and the use of physical controls such as sticky traps. Such methods aim to reduce pest populations while minimising disruption to the natural balance of the garden.

Regular monitoring and early intervention are the most effective pest control measures.

**Biological control**

Biological control If a crop is endangered despite your precautions, first consider biological controls. While most gardeners are familiar with the predaceous ladybug and praying mantis, many lesser known beneficial insects and even pathogens deserve recognition as effective natural controls.

*Long-tailed mealybug* (Pseudococcus longispinus). *These sap-sucking pests can cause stunting, leaf yellowing and reduced vigour, while honeydew deposits encourage the growth of sooty moulds.* (Image licensed from Shutterstock)

*Dragonfly, is a beneficial predator in the garden, feeding on flying insect pests such as aphids and fungus gnats.*
(Image Richard Bejsak)

## Natural predators

For many home gardeners, natural predators are the first line of defence against insect pests. Common beneficial species include ladybirds (*Cryptolaemus montrouzieri*), praying mantis (*Tenodera sinensis*), lacewings (*Chrysoperla carnea*), predatory mites (*Phytoseiulus persimilis*), parasitic wasps (*Leptomastix dactylopii*), and scale insect predators such as *Chilocorus* species.

Many of these biological control agents can be purchased from specialist suppliers. In some cases, they may also occur naturally in surrounding gardens and bushland.

Beneficial insects feed on pest species and do not damage plants. They are harmless to people and pets, and their populations decline naturally once pest numbers are reduced.

Encouraging beneficial insects helps maintain long-term pest balance in the garden. Providing suitable habitat and avoiding broad-spectrum sprays will support their activity and effectiveness.

*Native Australian ladybird beetle* (Harmonia conformis) *and larva. Both are beneficial predators that help control aphids and other sap-sucking pests.*
(Image Marilyn Steiner, reproduced with permission NSW DPI)

## Ladybirds

Ladybirds are effective predators at both the adult and larval stages. They feed on a wide range of pests, including aphids, scale insects, mealybugs, thrips and small caterpillars.

The mealybug destroyer (*Cryptolaemus montrouzieri*), a species of Australian origin, is one of the most widely used and successful biological control agents. It is particularly effective against mealybugs but will also feed on aphids and scale insects.

Ladybirds have a high feeding capacity, and even small numbers can significantly reduce pest populations. Once food sources decline, their numbers naturally decrease.

Larvae are often more voracious than adults and are a strong indicator of active control.

*Praying mantis* (Tenodera sinensis)*, a beneficial predator in the garden.*
(Image Richard Bejsak)

## Praying mantis

Praying mantises (*Tenodera sinensis*) are predatory insects that feed on a wide range of garden pests. They are easily recognised by their upright posture and folded forelegs, which they use to capture prey.

Mantises are generalist predators, feeding on insects such as caterpillars, flies and other soft-bodied pests. They typically hunt by remaining motionless and striking quickly when prey comes within reach.

Eggs are laid in protective cases (oothecae) attached to

vegetation. These cases hatch in warm conditions, releasing large numbers of young mantises.

While mantises can contribute to pest control, they are opportunistic feeders and will also consume beneficial insects. For this reason, they are best regarded as a supplementary control rather than a primary solution.

Mantises help reduce pest numbers but should not be relied upon as the sole control method.

*Lacewings* (Chrysoperla carnea) *are not only beautiful insects, but highly effective natural predators, helping control aphids and other soft-bodied pests in the garden.* (Image Richard Bejsak)

## Lacewing larvae

Lacewing larvae are highly effective predators, feeding on a wide range of small insect pests. Often referred to as 'aphid lions', they consume large numbers of aphids, as well as other soft-bodied insects such as whitefly and mites.

The larvae are active hunters and can significantly reduce pest populations during their development. However, they are also cannibalistic, particularly when food is scarce, so high densities are not always maintained.

Lacewing eggs are laid individually on the ends of fine stalks, which helps reduce cannibalism among newly hatched larvae. Adult lacewings are attracted to light and may be found near illuminated areas, from where they can be carefully collected and introduced into the garden.

Lacewing larvae are most effective when pest populations are already present.

*Lacewing larva ("aphid lion"), a voracious predator of aphids, whiteflies and other soft-bodied insect pests.* (Image Marilyn Steiner, reproduced with permission NSW DPI)

## Parasitic wasps

A range of small parasitic wasps play an important role in controlling insect pests in the garden. These insects are usually very small and often go unnoticed, but they are highly effective biological control agents.

Parasitic wasps lay their eggs in or on pest insects such as aphids, whitefly, scale insects, mealybugs asnd caterpillars. When the larvae hatch, they feed on the host, eventually killing it.

One of the most useful groups for home gardeners are aphid parasites, which attack and suppress aphid populations. Other parasitic wasps target whitefly and scale insects. Many species are extremely small and may only be seen with close inspection.

Parasitic wasps are harmless to people and pets and do not damage plants. Their presence is often indicated by parasitised pests, such as aphids that appear swollen or discoloured.

Parasitic wasps work best when pest populations are present but not yet severe.

*Tiny parasitoid wasp, one of several beneficial insects used in biological control programs.* (Image Marilyn Steiner, reproduced with permission NSW DPI)

*Predatory mite* Phytoseiulus persimilis, *a specialist predator used for biological control of two-spotted spider mite and other pest mites.* (Image Marilyn Steiner, reproduced with permission NSW DPI)

## Predatory mite

Also known as the Chilean mite, the predatory mite (*Phytoseiulus persimilis*) is widely used to control spider mites in greenhouse and outdoor crops. It is a highly active and voracious predator that can rapidly reduce, and in some cases eliminate, spider mite populations when conditions are favourable.

Introduce predatory mites early—they are most effective before spider mite populations become severe.

| Predator | Pest controlled |
|---|---|
| Ladybird | Aphids, mealybugs (including citrus mealybug), scale insects |
| Praying mantis | Ceneral insect pests |
| Dragonfly | Mosquitoes, gnats, flies and wasps |
| Lacewing larvae | Aphids, mealybugs, spider mites |
| Parasitic wasps | Aphids, whitefly, mealybugs |
| Predatory mite | Spider mites |

## Organic options

Organic pest control includes the use of low-toxicity sprays and simple, plant-safe treatments. One of the most widely used options is insecticidal soap, which works by coating insects and disrupting their ability to breathe. For best results, sprays should be applied directly to pests, particularly on the undersides of leaves. Because soaps can also affect beneficial insects, they are best used as a targeted treatment.

Seaweed (kelp) extracts can also be used as foliar sprays. While primarily applied to improve plant health and resilience, they may help reduce pest pressure, particularly in the case of spider mites. Care should be taken to avoid over-application, which can mark or stain foliage.

*Predatory mite* Neoseiulus cucumeris, *commonly used for biological control of thrips and other small insect pests.* (Image Marilyn Steiner, reproduced with permission NSW DPI)

A range of simple home-made sprays may also be used, typically based on soap solutions combined with plant extracts such as garlic or chilli. These mixtures can help deter pests but should be applied with care and tested on a small area first to avoid leaf damage.

Apply organic sprays early and directly to pests—repeated light treatments are more effective than heavy applications.

## Companion planting

Companion planting involves growing different plant species together so that one helps deter pests from another. Some plants produce strong scents or natural compounds that make the environment less attractive to insects, helping to reduce pest pressure.

In hydroponic systems, companion planting is less widely used than in soil gardens, but it can still be effective where space allows. Aromatic plants such as herbs are commonly used to help discourage pests.

The effectiveness of companion planting varies, and results will depend on the crops grown and local conditions. Practical experience is often the best guide, although suggested plant combinations are provided in the accompanying table.

Some companion plants may also attract beneficial insects, adding another layer of protection.

Others can improve growing conditions by providing light shading or acting as a physical barrier to pests.

Care should be taken to ensure that companion plants do not compete excessively for light, nutrients or space.

Companion planting can assist pest control but should be used alongside other management methods.

| Plant | Insect |
|---|---|
| Basil | Aphids, thrips, spider mite, |
| Mint | Cabbage worm |
| Sage | Cabbage worm |
| Rosemary | Cabbage worm |
| Onion | Spider mite, thrips, aphids |
| Garlic | Spider mite, |
| Radish | Stink bug, cu- cumber beetle |
| Marigold | Whitefly, aphids, nema- |
| Geranium | Leaf hopper, mosquitoes |
| Spearmint | Ants |
| Tansy | Ants, mosqui- |
| Pennyroyal | Ants, flies, mosquitoes |

## Sticky traps

Sticky traps are a simple and effective way to monitor and help control flying insect pests. They work by attracting insects—primarily through colour—and trapping them on an adhesive surface.

Yellow traps are widely used, as many pests such as whiteflies, aphids and fungus gnats are strongly attracted to this colour. Blue traps are often used for thrips. Traps should be placed at or just above plant height and checked regularly.

Sticky traps are most useful for early detection and monitoring. While they can reduce pest numbers, they are not usually sufficient as a sole control method and are best used alongside other strategies.

When pest problems arise, an effective management approach is to:

- Identify the pest and its natural enemies
- Monitor populations using sticky traps
- Assess the level of plant damage
- Apply appropriate control measures if thresholds are exceeded
- Evaluate the effectiveness of the action taken

*Yellow sticky traps attract and capture flying pests such as aphids, whiteflies, and fungus gnats, providing an effective tool for monitoring and control.*

*Algae thrive in moist, nutrient-rich conditions exposed to light, forming green growth that can compete with plants and clog hydroponic systems.*

## Algae control

Algae are common in hydroponic systems. While usually not directly harmful, excessive growth can compete for nutrients, reduce oxygen levels, and encourage pests such as fungus gnats.

Algae develop where light and moisture are present, so control is best achieved by excluding light from nutrient solutions and growing media. Channels, pipes and reservoirs should be kept light-proof using lids or opaque materials. In media systems, covering the surface can reduce growth, provided airflow to the root zone is maintained.

Good system hygiene is essential. Between crops, all containers, channels, irrigation lines and reservoirs should be cleaned and flushed. A mild disinfectant, such as diluted household bleach, may be used, followed by a thorough rinse.

Prevention is far easier than cure. A clean, well-maintained system with minimal light exposure will rarely suffer serious algae problems, and regular inspection allows early intervention.

## Plant Problems: Disease or Disorder

Not all plant damage is caused by disease. In hydroponic and greenhouse systems, symptoms such as leaf spotting, wilting, distortion, or poor fruit quality may arise from a wide range of causes—some biological, others environmental or cultural.

True plant diseases are caused by pathogens such as fungi, bacteria, and viruses. These often spread between plants and may progress rapidly under favourable conditions. Powdery mildew, bacterial wilt, canker, and mosaic viruses are typical examples, each with characteristic symptoms and patterns of development.

However, many common plant problems are not diseases at all. Disorders such as fruit cracking, poor pollination, pruning stress, herbicide injury, insect feeding damage, or guttation-related leaf marks can closely resemble disease symptoms. These issues do not spread from plant to plant, but instead result from environmental conditions, nutrient imbalances, or cultural practices.

*Lettuce mosaic virus, a seed-borne disease commonly spread by aphids, causing mottling, leaf distortion, and stunted growth.* (Image Dr William R. Jarvis)

Accurate identification is essential. Misdiagnosing a disorder as a disease can lead to unnecessary treatments, while overlooking a genuine infection may allow it to spread unchecked.

Careful observation—looking at the pattern of symptoms, the parts of the plant affected, and recent growing conditions—provides the best guide. Understanding the difference between disease and disorder is a key skill for every grower, and the first step toward effective control.and sound management.

## Common diseases

*Powdery mildew is a fungal disease, caused by* Podosphaera aphanis *or* Erysiphe necator. *It appears as white to ash-grey powdery spots on leaves, stems and fruit, leading to yellowing and leaf drop. It thrives in warm conditions with high humidity.* (Image Dr Mary M. Peet)

*Advanced downy mildew in cucumber. Lesions are typically bounded by leaf veins, creating the characteristic angular pattern that distinguishes the disease from many nutrient disorders, often leading to premature leaf death.* (Image licensed from Shutterstock)

*Mosaic virus infection in cucumber causing mottling, blistering and leaf distortion. Viral diseases are spread by aphids, contaminated tools and infected plant material.* (Image William R. Jarvis)

*Early blight in tomato causes dark leaf spots, yellowing and premature leaf death, often beginning on older foliage under warm, humid conditions.* (Image licensed from Shutterstock)

## Common diseases

*Fusarium is one of the most serious plant diseases in hydroponics and soil culture alike. Symptoms can vary considerably between crops and growing conditions, and the disease rarely appears the same twice.* (Image William R. Jarvis)

*Pythium root rot causes lettuce to collapse during warm weather. This image shows rapid wilting, leaf collapse and plant death typical of this water-borne disease in hydroponic systems.* (Image Len Tesoriero)

*Leaf spot diseases include fungal and bacterial pathogens causing brown or black spotting, yellow halos and premature leaf drop, particularly under warm, humid conditions.* (Image NSW DPI)

*Septoria leaf spot in tomato. Numerous dark lesions develop on older leaves, causing yellowing, premature defoliation and reduced plant vigour.* (Image licensed from Shutterstock)

## Common disorders (often mistaken for disease)

*Leafminer damage caused by larvae feeding between leaf surfaces, creating pale tunnels or blotches that reduce photosynthesis and plant vigour.* (Image Sue Korevaar)

*Sunscald and heat stress can cause bleached, papery leaf tissue and necrosis, particularly during periods of intense light, high temperatures or plant stress, reducing photosynthesis and crop productivity.*

*Fruit cracking is a physiological disorder caused by rapid water uptake after dry conditions, forcing the fruit to expand faster than the skin can accommodate.* (Dr Image Mary M. Peet)

*Thrips damage in lettuce appears as silvery, scarred or bronzed leaves with distorted growth. Thrips transmit viruses such as Tomato spotted wilt virus (TSWV).* (Image Dr Mary M. Peet)

## Common disorders (often mistaken for disease)

*Natural silver leaf patterning on pumpkin foliage is often mistaken for powdery mildew. Unlike fungal disease, the markings are uniform and part of normal growth.*

*Herbicide damage in tomatoes causes distorted growth, twisted leaves, cupping, stunting and fruit development following spray drift or contaminated compost.* (Image Dr Mary M. Peet)

*Catfacing causes misshapen, scarred and puckered fruit due to insect damage, poor pollination, temperature extremes, moisture fluctuations and stress during flowering.* (Image Dr Mary M. Peet)

*Sooty moulds are dark fungi that grow on honeydew excreted by sap-sucking insects such as whiteflies and aphids. They usually do not invade plant tissue.* (Image Dr Mary M. Peet)

# Maintaining your system

The efficient operation of a hydroponic system depends on regular maintenance and sound management. Allow 5–10 minutes each day to inspect the system for problems such as blockages, equipment wear or failure, insect pests, and early signs of disease or nutrient imbalance.

A common issue in NFT systems is restricted nutrient flow. This may be caused by blocked filters, algae in micro-irrigation lines, or roots growing into channel outlets. In some cases, roots may need to be trimmed to prevent ponding—the formation of stagnant pools within channels. Excessive *ponding* disrupts the nutrient film, reduces oxygen availability, and can lead to root rot and plant loss.

Systems using sand as a growing medium should be checked daily to prevent puddling. This occurs when water rises through the substrate and accumulates at the surface, keeping roots constantly saturated. Poor drainage is the usual cause. Improve drainage by incorporating coarse sand or gravel into finer media.

Between crops, the system should be thoroughly cleaned. Remove debris and scrub all components using a mild chlorine solution (household bleach). Circulate the solution through irrigation lines for up to two hours, then drain and flush the system with clean water. Before replanting, ensure the pump and filter are clean and functioning properly.

Test equipment such as pH and EC meters should also be maintained carefully. With proper handling and regular calibration, they will provide reliable measurements and long service life.

## Maximise growth

In recirculating hydroponic systems, plants draw nutrients from the solution, gradually altering its concentration, pH and EC. Without regular adjustment, growth will slow and plant health may decline. To maintain optimum performance, the nutrient solution should be replenished regularly or replaced when required.

As plants absorb different nutrients, the balance of the

solution shifts. Uptake of acidic elements will cause the solution to become more alkaline, while removal of alkaline salts will make it more acidic. As plants grow and develop, their nutrient demand increases, requiring careful adjustment of solution strength and composition.

The concentration of the nutrient solution is managed by adding fresh nutrients, while pH and EC are corrected using fresh water and appropriate buffer solutions.

Maximum growth can be maintained by following these simple daily steps:

**Top up the reservoir** with fresh water and circulate briefly to ensure even mixing.

**Check nutrient strength** (EC/TDS/cF):
Young leafy crops typically require 500–750 ppm, while mature fruiting plants range from 1000–2400 ppm, depending on the species and variety. Larger tomato varieties generally perform better at higher EC levels. Increasing EC can improve flavour, but may reduce fruit size. Adjust gradually using make-up solution, allowing time for thorough mixing. If EC is too high, dilute with fresh water.

**Check pH:**
Maintain a pH of approximately 5.5–6.0 for most crops. Adjust gradually using 'pH up' or 'pH down', avoiding concentrated additions to prevent overshooting.

Buffer solutions should be used sparingly, as they introduce additional nutrients—such as phosphorus, nitrogen or potassium—which may alter the balance of the solution. Most plants tolerate a moderate pH range, so frequent adjustment is not usually necessary.

## Calibrating meters

pH and EC 'pen-type' meters require periodic calibration to maintain accuracy. Each meter uses specific calibration solutions, available from hydroponic suppliers, aquarium stores and some hardware and garden centres. These are laboratory-grade solutions with a known, stable pH or conductivity.

For pH meters, common calibration points are pH 7.0 and pH 4.0. In most hydroponic systems, calibration at pH 7.0 is sufficient, as it is closest to the working range of typical nutrient solutions.

(Image Martin Caldeyro Stajano)

Conductivity meters are commonly calibrated using a solution of 2.76 mS/cm (approximately 1382 ppm on a TDS scale).

Calibration is straightforward. Rinse the probe with clean water to remove residue, then dry it. A dirty or wet probe can contaminate the calibration solution and affect accuracy. Immerse the probe in the calibration solution for about one minute to allow temperature compensation to stabilise, then check the reading.

If the meter is correctly calibrated, it should match the value of the solution. If not, adjust the meter according to the manufacturer's instructions—either using an automatic calibration function or manual adjustment.

Conductivity meters are calibrated in the same way.

## Keeping records

Good record-keeping is invaluable for the serious hydroponic gardener, allowing comparisons between crops and across seasons. Record any unusual observations, such as pest outbreaks affecting some plants but not others. Over time, this information can help refine practices, including companion planting and pest management.

When problems arise, they can often be traced back to a specific point in time. In many cases, plants take 7–10 days to show visible symptoms of stress or disorder. Accurate records help identify the cause, while corrective actions may take a further 7–10 days to produce a response.

For larger or more intensive systems, a simple financial record can also be useful. This should include capital costs, operating expenses (such as electricity and water), equipment maintenance, and consumables including nutrients and calibration solutions.

After each crop, summarise your records to highlight successes and identify areas for improvement. Consistent record-keeping allows you to fine-tune your techniques and maximise the productivity of your hydroponic system.

Hydroponics is both a science and a practical skill, built on careful observation and consistent attention to detail. Success does not depend on complexity, but on understanding the needs of plants and maintaining stable, well-managed growing conditions. By combining sound knowledge with regular monitoring and good system hygiene, even small-scale growers can achieve reliable, high-quality yields. With experience, confidence grows, and what begins as a simple system becomes a productive and rewarding way to grow fresh food year-round.

# Frequently asked questions

Hydroponics, or soilless culture, is the science of growing plants without soil. Water and nutrients are delivered directly to the roots, allowing plants to grow faster and reach harvest sooner. This method can produce consistent, high-quality crops that often outperform conventional growing systems. While the principles of hydroponics have been known for more than three centuries, it is only in recent decades that it has developed into a significant method of food production. Today, hydroponic produce is widely available through standard wholesale and retail channels. In this section, we answer ten of the most commonly asked questions about hydroponics.

## How nutritious are hydroponically grown vegetables and fruits?

Hydroponically grown fruits and vegetables have nutritional values comparable to those grown in soil, provided they are consumed soon after harvest. Like all fresh produce, nutritional quality declines over time.

Produce grown for supermarkets—whether hydroponic or soil-grown—gradually loses nutrients during transport, storage and display before reaching the consumer.

To help maintain freshness and nutritional value, some hydroponically grown crops are sold with their roots still attached, allowing them to remain alive and fresh until purchase.

## Do hydroponically grown fruit and vegetables taste better than soil-grown produce?

Hydroponically grown fruits and vegetables can taste just as good as those grown in soil when produced under optimal conditions. Flavour depends largely on how well the crop is managed rather than the growing method itself.

If plants are grown outside their ideal conditions, quality and flavour will decline. For example, tomatoes grown with incorrect nutrient strength (EC) may produce soft fruit with poor flavour. Maintaining proper growing conditions is essential for achieving the best taste.

## Do hydroponic plants mature faster than soil-grown plants?

Not all plants grow faster in hydroponic systems. Some, such as ferns and many Australian native species, show similar growth rates whether grown in soil or hydroponics, although hydroponic plants often display improved leaf size and overall appearance due to consistent nutrient supply.

In contrast, many vegetables, fruits and flowers do grow faster under hydroponic conditions—often by up to one-third—when managed correctly. This is because nutrients and water are delivered directly to the roots, eliminating competition and allowing plants to grow under optimised conditions.

## Are hydroponically grown fruits and vegetables larger than soil-grown produce?

Hydroponically grown fruits and vegetables are often larger and of higher quality when produced under optimal conditions. This is because plants receive a consistent, well-balanced supply of water and nutrients, supporting steady growth and development.

## Can all plants be grown in hydroponics?

Most plants can be grown hydroponically, although some are better suited to particular systems. Leafy crops such as lettuce and herbs perform well in water-based systems, while larger, fruiting plants are often better grown in substrate-based systems. An exception is fungi, such as mushrooms. Their nutritional requirements differ from those of plants, and they cannot be grown using standard hydroponic methods.

## Are inorganic nutrients harmful in hydroponics?

No. Hydroponic systems supply plants with the same inorganic elements they absorb from soil. In soil-based systems, organic materials such as compost and manure must first break down into these inorganic forms before plants can use them.

In hydroponics, these nutrients are delivered directly in a readily available form, allowing efficient uptake without altering their fundamental nature.

## How much nutrient solution do plants take up daily?

The amount of water and nutrients a plant takes up each day depends on its rate of photosynthesis and transpiration. These are influenced by factors such as leaf area, light intensity and duration, air temperature, humidity and air movement.

In general, larger plants use more solution than smaller ones. For example, a plant with five times the leaf area of a younger plant of the same variety will typically take up about five times more water and nutrients under similar conditions.

Environmental conditions also play a major role. Uptake is higher in hot, dry conditions and lower in cool, humid environments. Similarly, plants grown indoors, in shade or under cloudy conditions require less water and nutrients than those grown in full sunlight.

## Can the growing medium be re-used?

Most growing media can be re-used, provided they are clean and free from disease. If root diseases have occurred, the medium should be sterilised or replaced to prevent reinfection.

Some media, such as rockwool and coco peat, may even perform better in subsequent crops once they have been properly conditioned and stabilised.

## Is algae harmful to plants?

Algae will grow wherever light, moisture and nutrients are present. In most systems, it is not directly harmful if growth is limited and the system is functioning well.

Greater care is required in systems without a growing medium. Plant roots should be kept in darkness; the presence of algae in channels usually indicates excessive light exposure.

If algae develops, steps should be taken to exclude light from the system, as excessive growth can interfere with root health and system performance.

## Is hydroponics expensive to establish and run?

The cost of a hydroponic system depends largely on its level of complexity and automation. For home gardeners, simple systems can often be built using readily available materials, with basic components such as pumps and timers added at relatively low cost.

While nutrients represent an ongoing expense, they are generally affordable and used efficiently, making hydroponics economical to operate at a small scale.

## Is hydroponics just a trend?

No. Hydroponics is a well-established and practical method of food production, particularly valuable in regions facing population pressure, limited arable land or scarce water resources.

Compared with traditional soil-based farming, hydroponic systems can deliver faster crop cycles and, in many cases, reduced pest and disease pressure. They also offer environmental benefits, particularly through efficient water use and reduced waste.

## How difficult is hydroponics for a beginner?

Hydroponics is not difficult for beginners, provided a simple system is used and basic principles are followed. While the method may appear technical at first, it is largely a matter of routine observation and regular maintenance.

Most home systems require only a few minutes each day to check water levels, nutrient strength and plant health. Once established, the process becomes familiar and straightforward.

Starting with an easy-to-manage system and reliable crops, such as leafy greens or herbs, will build confidence quickly. As experience grows, more complex systems and a wider range of plants can be introduced.

With a little care and consistency, hydroponics is an accessible and rewarding way to grow fresh produce at home.

## What is the easiest system to start with?

For beginners, the simplest systems are those with few moving parts and minimal maintenance. Wick systems and basic water culture (deep water culture) are among the easiest to set up and operate, as they require no pumps or only simple aeration. These systems are well suited to leafy crops such as lettuce and herbs.

Small nutrient film technique (NFT) systems can also be used by beginners, provided they are kept simple and monitored regularly. While they require a pump, they offer efficient nutrient delivery and good plant growth when properly maintained.

For those who prefer a more forgiving system, media-based methods using materials such as coco peat or expanded clay are a good option. These provide support for the roots and buffer against fluctuations in moisture and nutrients.

Starting with a simple, reliable system allows beginners to gain confidence and understand the basics before moving on to more advanced designs.

## What happens if the power fails?

In systems that rely on pumps or aeration, a power failure can quickly disrupt water flow and oxygen supply to the roots. In NFT and deep water culture systems, this may lead to plant stress if the outage is prolonged.

Short interruptions are usually not serious, but extended outages can cause root damage or plant loss. To reduce risk, ensure good system design, avoid overloading channels, and consider backup options such as battery-powered air pumps or a generator.

## Can hydroponics be organic?

The term "organic" is generally associated with soil-based growing systems that rely on natural inputs such as composts and manures. Traditional hydroponics, by contrast, uses mineral nutrient solutions, so it is not usually classified as organic under most certification standards.

However, it is possible to apply organic principles in hydroponic systems by using natural inputs and avoiding synthetic chemicals. This approach—often referred to as organic hydroponics or bioponics—relies on organic nutrient sources that must be broken down into plant-available forms, often with the aid of beneficial microbes.

While effective, organic hydroponics is more complex to manage than conventional systems, as maintaining nutrient balance and solution stability can be challenging.

For most home growers, standard hydroponic methods offer a simpler and more reliable approach, while still producing clean, high-quality crops with minimal chemical use.

## Can I grow hydroponically indoors?

Yes. Hydroponic systems are well suited to indoor growing, provided plants receive adequate light and ventilation. In the absence of natural sunlight, artificial lighting—such as LED grow lights—can be used to supply the energy required for photosynthesis.

Indoor systems offer a high level of control over growing conditions, including temperature, humidity and light duration. This can lead to consistent growth and year-round production, regardless of outdoor weather.

Good airflow is important to prevent excess humidity and reduce the risk of disease. Simple measures such as using a small fan and avoiding overcrowding will help maintain a healthy growing environment. With appropriate lighting and basic environmental control, hydroponics can be successfully practised in a wide range of indoor spaces, from kitchens and spare rooms to dedicated growing areas.

# Glossary

## Aeration

The process of supplying oxygen to the root zone, usually by introducing air into the nutrient solution. Adequate aeration is essential in water-based systems to prevent root stress and promote healthy growth.

## Aeroponics

A system in which plant roots are suspended in air and continuously or intermittently misted with a fine spray of nutrient solution. While used by hobbyists, aeroponics accounts for only a very small proportion of global hydroponic production, with limited commercial use.

## Aquaponics

A system that combines hydroponics with aquaculture (fish culture), in which nutrient-rich water from fish tanks is used to supply plants. Waste produced by the fish is converted by beneficial bacteria into nutrients that plants can absorb. The plants, in turn, help to filter and clean the water, which is recirculated back to the fish. Aquaponics operates as a balanced, closed-loop system requiring careful management of both plant and fish health.

## Conductivity Factor (CF)

An indicator of the strength of a nutrient solution, measured using a CF meter. This term is commonly used in the UK, New Zealand and by some growers in Australia. The unit of measurement is the CF unit. (See also Electrical Conductivity – EC.)

## Deep water culture (DWC)

A hydroponic system in which plant roots are suspended in a nutrient solution, typically within a container or tank. An air pump and air stone are used to provide oxygen to the roots. DWC is simple to operate and well suited to small-scale and beginner systems.

## Drip system

A hydroponic system in which nutrient solution is delivered directly to the base of each plant through small emitters (drippers). Excess solution may either drain away (open system) or be collected and reused (closed system). Drip systems are widely used for larger plants and commercial production.

## Electrical Conductivity (EC)

A measure of the total concentration of dissolved salts in a nutrient solution, determined using an EC meter. This is the standard term used worldwide. It is commonly expressed in milliSiemens per centimetre (mS/cm) or microSiemens per centimetre (μS/cm).

EC indicates overall solution strength but does not identify individual nutrient components. (See also Conductivity Factor – CF.)

*Unit relationships:* 1 mS/cm = 10 CF units = 1000 μS/cm

## Fertiliser

Chemical compounds, usually in solid or liquid form, that dissolve in water to form part of a nutrient solution. These supply essential elements required for plant growth—for example, potassium nitrate ($KNO_3$). A wide range of pre-mixed hydroponic fertilisers is available, particularly for home growers, offering balanced formulations for different crops and growth stages.

## Flood-and-Drain (Ebb-and-Flow)

A recirculating hydroponic system in which nutrient solution is periodically pumped into the growing area and then allowed to drain back to a reservoir. Plants are typically grown in a medium and irrigated from below (sub-irrigation).

Early systems, including those developed by the US Army after World War II, used large gravel beds. Today, a wide range of growing media is used. While less common in large-scale commercial production, flood-and-drain systems remain popular for propagation, pot plants and small-scale growing due to their simplicity and reliability.

## Growing Medium

Materials used to support plant roots and provide a balance of water, nutrients and air for healthy growth. Growing media are commonly placed in containers such as pots, bags or trays and supplied with nutrient solution via drippers or sub-irrigation systems. Growing media are broadly classified as organic or inorganic:

**Organic:** bark, peat, sawdust and coir (coco peat). Coir has become widely used due to its sustainability and favourable water-holding properties.

**Inorganic:** expanded clay, gravel, perlite, pumice, rockwool (horticultural), sand (quartz) and scoria.

Many growers use blends of different media to optimise drainage, aeration and moisture retention for specific crops and systems.

## Hydroponics

A method of intensive horticulture in which plants are grown without soil, using nutrient-rich water solutions. Plants may be grown either in a water-based system or in an inert growing medium, with all essential water and nutrients supplied directly to the roots.

The term hydroponics is widely used in Australia, New Zealand and the United States, while soilless culture is often used more broadly worldwide. In some contexts, true hydroponics refers specifically to water-based systems, distinguishing them from media-based methods.

## Ions

When fertilisers dissolve in water, they separate into electrically charged particles known as ions. These include positively charged ions (cations), such as calcium ($Ca^{2+}$), and negatively charged ions (anions), such as nitrate ($NO_3^-$). The balance of these ions in solution is essential for proper nutrient uptake by plants.

## NFT (Nutrient Film Technique)

A hydroponic system in which a shallow stream of nutrient solution flows continuously through channels or gullies, bathing the plant roots before being recirculated. The solution is maintained as a thin film to ensure adequate oxygen supply to the roots.

NFT is widely used for leafy crops such as lettuce and herbs, particularly in small-channel systems, due to its efficiency and ease of management.

## Non-recirculating systems

Media-based systems in which plants are supplied with fresh nutrient solution, with excess solution draining away rather than being reused. These are also known as open or free-drainage systems.

Traditionally, most media systems operated this way, but there is an increasing shift toward recirculating systems. In these, runoff is collected, treated if necessary, adjusted for nutrient balance, and returned to the system for reuse.

## Nutrients

The essential mineral elements required for plant growth, broadly comparable to vitamins in human nutrition. These are divided into two groups:

**Macronutrients (major nutrients):** required in larger amounts—nitrogen (N), potassium (K), calcium (Ca), phosphorus (P), magnesium (Mg) and sulphur (S).

**Micronutrients (trace elements):** required in small quantities—iron (Fe), manganese (Mn), zinc (Zn), boron (B), copper (Cu), chlorine (Cl) and molybdenum (Mo).
While plants may absorb other elements, these are the primary nutrients essential for healthy growth.

## Osmosis

The natural movement of water across a semi-permeable membrane, such as plant root cells. Water moves from a weaker (less concentrated) solution to a stronger (more concentrated) solution, helping to balance concentrations on both sides of the membrane.

## Nutrient solution

A water-based mixture containing dissolved mineral nutrients required for plant growth. It supplies all essential elements directly to the roots and is the primary source of nutrition in hydroponic systems.

## pH

A measure of acidity or alkalinity in a solution, determined using a pH meter or colourmetric strips. It has no units. A pH of 7.0 is neutral; values below 7.0 are acidic and those above 7.0 are alkaline. Adding acid lowers pH, while adding an alkali raises it.

## Photosynthesis

The process by which plants use light energy to convert carbon dioxide ($CO_2$) from the air and water ($H_2O$) from the roots into sugars, which provide energy for growth. Oxygen ($O_2$) is produced as a by-product and released into the atmosphere.

## Pollination

The transfer of pollen from the male anther to the female stigma of a flower. This process is essential for fertilisation and the production of fruit and seeds in many plants.

## ppm (parts per million)

A unit used to express the concentration of a substance in a solution, commonly used for nutrients in hydroponics. One ppm is equivalent to 1 milligram per litre (mg/L).

It refers to the concentration of the actual nutrient element, not the fertiliser supplying it. For example, if a fertiliser contains 20% of a nutrient, 5 grams of fertiliser are required to supply 1 ppm of that nutrient in 1000 litres of water.

## Recirculating systems

Systems in which nutrient solution is collected, reused and circulated repeatedly. These are also known as closed systems.

Some systems, such as NFT, operate continuously, while others, such as flood-and-drain, run intermittently. In media-based systems, runoff may be collected, treated if necessary, and returned to the system.

## Reverse osmosis (RO)

A water purification process in which pressure is applied to force water through a semi-permeable membrane, removing dissolved salts and impurities. This produces high-quality, low-mineral water suitable for preparing nutrient solutions.

## Reservoir

A container used to store nutrient solution within a hydroponic system. It supplies solution to the plants and is typically fitted with a pump in recirculating systems. The reservoir should be kept clean and protected from light to maintain solution quality.

## Rockwool

A growing medium made by melting volcanic rock and spinning it into fine fibres, then forming it into blocks, cubes or slabs. It is widely used in hydroponics for both propagation and crop production due to its excellent water-holding capacity and aeration. Its use is particularly common in regions with local manufacturing.

## Root zone

The area surrounding the plant roots where water, nutrients and oxygen are absorbed. Maintaining proper conditions in the root zone is essential for healthy growth and optimal plant performance.

## Runoff

Excess nutrient solution that drains from the growing medium after irrigation. In recirculating systems, runoff may be collected, treated if necessary, and reused; in open systems, it is discarded, sometimes referred to as *free drainage*.

## Soilless culture

The cultivation of plants without soil, using water-based systems or inert growing media to supply all required nutrients. This term is widely used internationally and is synonymous with hydroponics.

## Total Dissolved Solids (TDS)

The total concentration of dissolved inorganic substances in a solution, typically expressed in parts per million (ppm) or milligrams per litre (mg/L).

Meters labelled as TDS meters do not measure dissolved solids directly; instead, they estimate TDS by converting electrical conductivity (EC) readings using a conversion factor. As a result, readings may vary in accuracy depending on the calibration used. (*See also Electrical Conductivity – EC.*)

## Transpiration

The movement of water through a plant, replacing moisture lost through evaporation from the leaves. This process helps regulate temperature and maintain nutrient flow within the plant.

## Truss

A compact cluster of flowers or fruit borne on a single stem, as seen in crops such as tomatoes.

# Acknowledgements and Image Credits

Certain images reproduced in this publication are used for educational, research and illustrative purposes. Every effort has been made to identify, acknowledge and obtain permission from copyright holders where required.

Unless otherwise stated, all photographs and illustrations are by the author or are in the public domain. Images reproduced from other sources are credited in the relevant captions and are used with permission or under licence.

Image sources include the late Marilyn Steiner (reproduced with permission from the NSW Department of Primary Industries), Haifa Group, Richard Bejsak, Dr Mary M. Peet, Dr Mike Nichols, Bruce Christie, the late Dr William R. Jarvis, Dr Len Tesoriero, Eyal Ronen and Martin Caldeyro Stajano. The publisher gratefully acknowledges these organisations and individuals for their contributions.

Any copyright owner who has not been properly identified or credited is invited to contact the publisher so that appropriate corrections or acknowledgements may be incorporated in future editions.

# About the author

Steven Carruthers is an award-winning horticultural journalist and former publisher and editor of *Practical Hydroponics & Greenhouses*, a multi-award-winning magazine he led for 27 years. Through his work, he has contributed extensively to the development and communication of hydroponic growing systems for both commercial and home growers.

He has a long-standing interest in sustainable food production, food security and environmental management, with a focus on practical, accessible solutions for small-scale growers.

Steven is also a former Royal Australian Navy serviceman and commercial diver, and the author of several best-selling works on Australia's wartime history, including *Australia Under Siege*, *A Parting Shot*, *Ghost Below: The Lost Submarine of Sydney Harbour*, and *Japanese Submarine Raiders*.

# Index

# NOTES

www.ingramcontent.com/pod-product-compliance
Ingram Content Group UK Ltd.
Pitfield, Milton Keynes, MK11 3LW, UK
UKHW060011051026
15591UKWH00002B/139
* 9 7 8 1 7 6 4 7 4 6 4 0 3 *